SITE INVESTIGATIONS AND FOUNDATIONS EXPLAINED

SITE INVESTIGATIONS AND FOUNDATIONS EXPLAINED

M. Carter
and
M.V. Symons
School of Engineering
University of Wales College of Cardiff

PENTECH PRESS
London

First published 1989 by
Pentech Press Limited
Graham Lodge, Graham Road
London NW4 3DG

British Library Cataloguing in Publication Data
Carter M.
Site investigations and foundations explained: a practical guide for civil engineers, architects and builders.
1. Buildings. Construction. Sites
I. Title II. Symons, M.V.
690

ISBN 0-7273-1907-8

Printed and bound in Great Britain by Billing & Sons Ltd, Worcester and London

CONTENTS

PREFACE ix

PART I: SETTING THE SCENE

CHAPTER 1 — THE OVERALL VIEW **1**

Where the site investigation and its findings fit into the planning, design and construction processes 1

What the geotechnical engineer must consider and evaluate 4

What he needs to know about the project 8

How he sets about appraising the site; the site investigation 9

How he interprets the site investigation results and communicates his findings; the site investigation report 10

Use of the site investigation report by those involved in the planning, design and construction of projects 12

CHAPTER 2 — TYPES OF FOUNDATIONS AND EARTH RETAINING WALLS **14**

The importance of foundation considerations in the majority of site investigations 14

Review of foundation types, with discussion of the ground conditions and building types to which they apply 16

Review of earth retaining wall types 33

PART II: THE TYPICAL SITE

CHAPTER 3 — THE SITE INVESTIGATION 36

Specifying what is required from the investigation, phasing and organisation of the work 36

The desk study; use of maps, records and aerial photographs 38

Appraising the site and planning the ground investigation: appropriate methods and the number, positions and depths of boreholes 47

Pits and borings; common methods 53

Sampling and field testing: suitable frequency and outline procedures 59

Geophysical methods appropriate to civil engineering 72

CHAPTER 4 — LABORATORY TESTS 77

Applicability of tests to design and soil conditions; which tests give what information 77

Outline procedures for common tests, with discussion of problems and misleading results that can arise 80

CHAPTER 5 — THE GEOTECHNICAL REPORT 105

Contents and layout of a typical report: factual and interpretive reports 105

Soil descriptions and classification systems. Borehole logs. Rock descriptions 108

Tests results and recommendations given in reports: presentation and terms used 124

CHAPTER 6 — GETTING THE MOST OUT OF THE GEOTECHNICAL INFORMATION 131

How the geotechnical engineer reaches his conclusions and recommendations: reviewing the information 131

Problems due to lack of knowledge 140

Looking for warning signs which could indicate future problems 141

Making further use of the report at the detail design stage 147

PART III: SITES WITH SPECIAL PROBLEMS

CHAPTER 7 — PROBLEM SOILS **149**

Soil types that give rise to particular problems: identification of problem soils and precautionary measures 149

CHAPTER 8 — MINING AREAS **162**

Past mining; the problems caused by void migration, localised and general rock collapse 162

The importance of a correct desk study: details of correct procedures 169

The special requirements of site investigations in old coal mining areas 179

Precautionary and remedial measures for construction in old coal mining areas 184

Current and future mining. Precautionary and preventive measures 197

CHAPTER 9 — DISTRESS IN EXISTING STRUCTURES **207**

Possible causes of cracks and movement in existing buildings; the problems of diagnosis. The long term consequences and possible remedial measures 207

PART IV: A DEEPER LOOK

CHAPTER 10 — PRINCIPLES OF SOIL MECHANICS **216**

Idealised soil structure 216

Shear strength: total and effective stresses. The distinction between the immediate and long term response of soil to loads, and its significance for geotechnical engineering design 219

Consolidation properties of soil 228

CHAPTER 11 — FOUNDATION DESIGN **233**

How foundations fail: ultimate and allowable bearing capacity 233

Dealing with eccentric loading and variable ground conditions 241
Calculations of stresses and displacement by elastic theory: use of influence factors 245
Estimation of magnitude and rate of consolidation settlement. Differential settlement. Rapid methods of assessment. Settlement of sands. Tolerable settlement 254
Piled foundations; ultimate and working loads of piles and pile groups. Negative skin friction 263
Pile driving formulas and pile load tests 273

CHAPTER 12 — EARTH RETAINING WALLS AND SLOPES 284

Earth pressures on walls 284
Design rules for rigid walls, braced excavations and flexible walls 293
Stability calculations for flexible walls 298
Stability problems in natural slopes, cuttings and embankments 307
Modes of slope failure and review of types of analysis used. Slope stability curves 309

REFERENCES AND BIBLIOGRAPHY 324

INDEX 328

PREFACE

In this book we have set out the various aspects of geotechnical engineering, from the planning and organisation of site investigations, site work and soil testing methods, through to the appraisal of site investigation reports and their use in foundation design. It has been written to meet the requirements of a wide variety of construction professionals, including civil, structural and geotechnical engineers, architects and planners. We have tried to give clear explanations which we hope the non-specialist will find informative and useful, and to bridge the gap between theory and practice, which we feel will be particularly helpful to those taking up careers as geotechnical specialists. To achieve these twin aims, the book has been structured so that readers can pick out particular topics that interest them and can follow each topic to the depth they require.

The work is divided into four main sections, each giving a greater degree of specialisation and complexity than the previous. "Setting the scene" puts the site investigation work in the context of the construction project, and describes what the geotechnical engineer must consider and how the results of the investigation may be used by other professionals. There is also a review of foundation types and their applicability. "The typical site" describes common site investigation and laboratory testing methods and includes advice on the organisation of investigations, on how to select the most appropriate methods and on the amount of boring, sampling and testing required. The contents of a typical geotechnical report are discussed, with explanations of specialist terms used and emphasis on how to obtain the maximum amount of information from it. "Sites with special problems" deals with problem soils, such as fills and expansive clays, and with the difficulties of identifying causes of distress in existing buildings. The problems of mining areas and the special requirements of site investigations in them are dealt with in some detail. "A deeper look" is for those who wish to delve into soil mechanics and geotechnical engineering design, and includes explanations of why particular design approaches are adopted.

Chapter 1

THE OVERALL VIEW

WHERE THE SITE INVESTIGATION FITS INTO THE PLANNING, DESIGN AND CONSTRUCTION PROCESSES

The importance of the ground conditions at a particular site and the influence they have on a project depend on a number of factors, such as the type of structure being proposed, the sensitivity of the structure or any installed machinery to ground movements, the size of the project and the types of soils underlying the site. For example, for the design of a single house, the plot will usually have already been chosen, the design will be governed by the wishes of the architect or owner and the foundation design will normally be standard, unless ground conditions are exceptional. Poor ground conditions are normally accommodated by deepening the foundations or using short bored piles, with no change in the design or construction methods except for the foundations. Where ground conditions are exceptionally poor they may have a much greater influence on the choice of structure; timber frame construction being chosen to save weight instead of traditional brick and block construction, for instance.

With larger buildings such as factory units, which are often built on poor ground, an early consideration of ground conditions with careful choice of light, flexible structures, can have a considerable influence on initial costs and long-term serviceability. With larger housing schemes, soil conditions may affect the whole layout of the scheme. Parks and recreational areas can be located where ground conditions are worst so that the cost of foundations and access roads can be minimised. With large and complex civil engineering projects, the ground conditions may affect not just the cost, but the location and even the feasibility of a project. This is especially true of schemes which are sensitive to ground movements, such as nuclear power stations, or those with a high geotechnical content such as roads, dams and tunnels.

Consideration of the ground conditions at a site and an appreciation of their significance to a proposed project is important at all stages of planning, overall design, detail design and construction.

The example of the housing scheme given above illustrates how early attention to site conditions can avoid design problems at a later stage. Power stations and dams are obvious examples of projects where ground conditions are important at an early stage but for all projects, even the single house, neglect of site conditions early on can have disastrous consequences. How many people have bought land to build their dream homes only to find that the site is covered by loose fill, is on an unstable hillside, or is underlain by very soft clay?

At the design stage, careful location of structures and choice of a suitable structure and finishes can help reduce foundation costs and the incidence of cracking and distress during the life of the building. It is at this stage in particular that an appreciation of the significance of how the ground conditions will affect the structure is most important. The foundation types available, their effectiveness in adequately supporting a structure and their costs should be considered as an integral part of the design. Too often designs for the main structure are prepared with little or no consideration of the ground conditions and are then passed over for the geotechnical engineer to add the foundations, almost as an afterthought. In poor ground conditions it may be impossible to design against excessive settlement or foundation failure of a heavy structure at reasonable cost and a better solution may be to design a lighter structure or one which can tolerate greater settlement. In many cases, designs virtually ignore the soil conditions: shallow foundations are designed for buildings on expansive soils, leading to cracking; and floor slabs are laid directly on highly compressible clays so that they distort and part company with the walls as the building settles.

For the majority of projects concerned with building construction, the ground conditions influence the initial stages of the work, when site access is being prepared and foundations are being constructed. Will trucks and plant be able to travel over the site without transforming it into a quagmire or must hardcore be provided or the site stabilised with lime? Will excavations be stable and, if not, what sort of support will be required? Will tracked or wheeled plant be best? Will there be groundwater problems in the excavations and can these be avoided by constructing foundations in the summer months? If pumping from large excavations is required, what pumping capacity should be installed? Is the ground especially susceptible to heave and softening at the base of foundations and, if so, should work be programmed so that the process of excavation, foundation construction and backfilling takes place as rapidly as possible?

Good working practices avoid most of these problems on smaller projects. The project is, hopefully, planned to begin in the spring to take advantage of the drier ground conditions in the initial phases. Excavations are kept well drained and concreting is begun as soon as excavation is completed to reduce the risk of wetting or drying of the base of the excavation with consequent shrinkage/swell problems. On road construction projects, the subgrade is protected by a layer of earth or by laying sub-base as soon as earthworks are completed up to formation level.

It is when construction begins that deficiencies in the site investigation begin to show up, causing headaches for the site staff, irritation and expense to the client and opportunity for the claims expert. It is a source of unceasing wonderment how all those boreholes could be so located that they totally missed the buried river channel through the centre of the site, or the peat deposits lying beneath so much of it. This is often seen as a failure of the site investigation, and in a way it is. Such failures are often laid at the door of the site investigation engineer and it may indeed be his ineptness at noticing and interpreting surface features which resulted in a poor choice of borehole locations. On the other hand, the client who thinks that site investigations are a waste of money and that a single borehole is quite sufficient to reveal the subsoil conditions must expect some nasty surprises.

Most of the shortcomings of site investigations are not the result of any particular failure but arise simply because of the nature of the problem. Consider, for instance, a 10-storey concrete building, 40m square. This might contain 4000m^3 of concrete but the volume of soil beneath the foundations which would be involved in settlement or potential failure would typically be 40m x 40m x 25m deep; some 40000m^3. The concrete, which is a carefully controlled, man-made material, might be tested every 50m^3, giving 800 samples tested. The site investigation might specify 6 boreholes to 30m depth, with, at most, 10 samples tested from each borehole. Thus, only 60 samples would be tested, or roughly one sample for every 700m^3, even though soil is a natural substance with a much greater variability than the concrete.

WHAT THE GEOTECHNICAL ENGINEER MUST CONSIDER AND EVALUATE

The first and most obvious consideration for any structure is, "Will it stand up?". Translated into the geotechnical engineer's terms, this becomes, "What is the ultimate bearing capacity of the ground?". Thus, soil strength is an early and important consideration.

An initial estimate of the allowable bearing pressure is usually obtained by dividing the ultimate bearing capacity (the bearing pressure at failure) by an appropriate factor of safety, typically 2.5 or 3. To the structural engineer, this may sound exceptionally conservative but there are good reasons for it. First, the ground conditions are much more variable than is the case with manufactured materials and, even after a thorough site investigation, remain largely unmeasured. Second, soil-structure interaction systems are complex, highly redundant and difficult or impossible to analyse accurately. Third, the stress-strain behaviour of soil is non-linear and stresses close to failure are accompanied by large plastic deformations and possibly by long-term creep movements. The high factor of safety keeps stresses well below this region.

The choice of a high factor of safety against bearing capacity failure means that, for many foundations on clay soils, settlement is not a problem. However, settlements are an overriding consideration in a significant number of cases, particularly with large foundations or soft clays. Because of this, consolidation tests should always be carried out so that settlements can be calculated. If the allowable bearing pressure based on ultimate bearing capacity would cause unacceptably high settlements then the value finally recommended must be reduced and be based on settlement considerations instead. What is acceptable depends on the structure: a maximum value commonly used for conventional rigid structures is 25mm (or an old-fashioned inch). Settlement estimates often receive less attention than they deserve (or even none at all), with the result that excessive settlements cause far more problems than do bearing capacity failures.

Sands and gravels present special problems of their own. Because the particles do not stick together, as clays do, it is difficult or impossible to obtain undisturbed samples. In addition, very coarse soils may contain pieces that are simply too big to be sampled. This is overcome by the use of insitu testing, usually in the form of a metal cone or rod which is driven or pushed into the soil. Estimates of ultimate bearing capacity and settlement are obtained from these tests. The ultimate bearing capacity of foundations on granular soils is very high, except for narrow foundations on loose sands, and, at the design stage, settlement is usually the overriding consideration.

Many problems caused by ground movements are not associated with consolidation settlements at all but with the tendency of some clays to shrink and swell as the moisture content changes. The usual way to overcome this is to take the foundations below the zone of seasonal moisture content changes. Ground floor slabs are designed to accommodate movements or suspended floors are used. This is a problem which is often overlooked with the result that specified foundation depths are much too shallow. Even when faced, it can be a source of irritation between the geotechnical engineer and the structural engineer because the solution lies in the design of both the foundations and some of the structural details.

Groundwater levels are important since they affect both the ultimate bearing capacity and the settlement of foundations. High water levels can result in flooding of foundation excavations or instability of the sides. In certain soils a high water table can cause heave or even a quicksand effect in the base of excavations. These problems can often be overcome by keeping foundations shallow, by providing a simple pump or by programming work so that foundations are built during the summer months when groundwater levels are likely to be lower. At the other extreme, expensive dewatering or sheet piling driven below the base of excavations may be necessary. For certain applications, caissons may be a solution to the problems of weak soils with a high water table.

Seasonal variations in groundwater levels must be taken into account when considering the groundwater levels encountered during the site investigation. Not only are seasonal variations important when designing the foundations and planning the construction program, but for some clays the resulting seasonal shrinkage and swelling can cause major cracking problems to the structure.

Variability of site conditions is a major consideration. Can the site be treated as one area from the foundation design point of view? How many trial holes will be needed and where should they be located? Should some areas of the site be more thoroughly investigated because of particular problems in those areas? In some ways this is a chicken-and-egg problem because the geotechnical engineer must make an assessment of subsoil conditions before he can plan the ground investigation but until he has the results of this investigation he cannot know what the subsoil conditions are. This is partly overcome by first carrying out a desk study to obtain as much written information as possible about ground conditions in the area, and by excavating preliminary trial holes to give an overall picture of subsurface conditions at the site before carrying out a more detailed investigation. Previous use of the site may have important consequences and the desk study should also provide information on this particular aspect. On many sites, subtle changes such as slope variations, vegetation type and land use reflect variations in subsoil conditions. The experienced engineer will be on the lookout for such clues when he visits the site and should be able to use them to plan the site investigation so that it will yield the maximum information for the minimum cost.

The project itself is another obvious consideration. Does it have a major geotechnical component, such as a road, dam, retaining wall or tunnel? Is it a building and, if so, what type of foundations is it likely to have? Where are structures likely to be located? Here again, we have the chicken-and-egg situation. Whenever possible, the site investigation should be carried out so that changes in project layout do not result in structures being positioned where there has been an embarrassing lack of ground investigation work.

In planning the site investigation work, the engineer should start by considering the proposed project. He should consider what geotechnical problems will arise and what information will be needed to solve these. Does he need to give advice on foundation types, bearing capacities and depths? Should the stabilities of slopes and excavations be considered? Is there a seepage problem? Knowing the problems to be solved, he can then decide on what soil parameters need to be measured. This defines the types of tests required and whether testing should be on disturbed or undisturbed laboratory samples or carried out insitu. He will usually have some idea of what types of soil or rock are likely to be encountered and can then select the drilling, sampling and site testing methods that are appropriate to the anticipated ground conditions and that will give the types of samples and measurements he needs.

The letting of the contract can cause difficulties at the early planning stage. This is because the contract, normally carried out by a specialist site investigation contractor, must include a specification on such things as the types of drilling, sampling and testing to be used, with a bill of quantities showing a tentative amount for each item. This implies a specialist knowledge on the part of the client, who is likely to be a consulting engineer or a central or local government agency. For a large consultant or government agency with its own team of geotechnical specialists this may present no great difficulty but, for a small firm of architects or structural engineers, knowing what to specify so that maximum benefit is obtained from the site investigation is not easy.

WHAT THE GEOTECHNICAL ENGINEER NEEDS TO KNOW ABOUT THE PROJECT

The short answer is, "as much as possible". The purpose of the project, the types of buildings, their heights, methods of construction, loading and sensitivity to ground movements are all important. The use to which buildings are to be put and any machinery they will contain may have a marked effect on the settlements that can be tolerated. The proposed layout of the development will affect where the site investigation effort is concentrated. This may not be known in the early stages, or it may be affected by the site investigation findings. If there is any possibility that the layout will be changed, the site investigation should aim to cover all alternatives. It is embarrassing to have to report that, as a result of changes to the proposed layout, there is no geotechnical information available for the foundation design of many structures and that most of the costly information obtained relates to areas which are now to be empty space.

If the project is to include access roads, the appropriate tests must be carried out, but it is not necessary to know details of the traffic expected unless a pavement thickness design is called for. If there are to be yards used for heavy storage, the expected loadings must be estimated so that the amount of settlement can be anticipated. This is especially important where rail-mounted cranes are used.

When deciding on the depths to take trial pits, boreholes or probes, the engineer must consider what type of foundations are likely to be used. The site investigation must extend to below the depth that will be affected by the foundation pressures, and deep foundations will obviously require a deeper site investigation. A deep basement may be specified for non-geotechnical reasons: to provide storage or parking for instance. Caissons may be considered early on because they are easier to construct. In such cases, the geotechnical engineer must be informed and consulted on the choice of foundation types. The choice between shallow foundations or piles will affect the depth of soil that is stressed so affecting the depth to which the trial holes must be taken. But the choice of foundation may itself depend on the results of the site investigation. In such cases, the geotechnical engineer must use his judgement when deciding how deep to investigate ground conditions. When in doubt it is better to err on the side of caution and take at least some of the borings deeper than the normal minimum requirement. Often, these problems can be overcome simply by carrying out a good desk study with a reconnaissance of the site and surrounding areas. This should give an indication of the likely ground conditions and may provide information on the types of foundations used for existing structures in the area.

THE SITE INVESTIGATION: HOW THE GEOTECHNICAL ENGINEER SETS ABOUT APPRAISING THE SITE

Having received a brief giving the location of the site and an outline of the proposed development, the geotechnical engineer's first priorities are to gather what information he can about ground conditions in the area and to visit the site. This is the reconnaissance phase of an investigation. It is not necessary for a full desk study to be completed before the visit, but he should at least have consulted a geological map of the area.

Once the site has been visited and the full desk study completed, he should have a good idea of the most appropriate methods to use and be able to estimate the extent of site investigation work required. He can now plan the site investigation in detail.

The next phase is the site investigation proper using the appropriate choice of trial pits, boreholes and probes. The number and types of samples, insitu tests and laboratory tests will depend on the nature of the project and the subsurface conditions encountered. Once samples have been obtained they will be sent to the laboratory and an appropriate program of testing drawn up.

On small or straightforward projects the field work takes place as a single phase, but for more complex projects or difficult ground conditions it is often convenient to split the work into two phases. In the preliminary phase, relatively few trial holes or probes are used and the aim of the investigation is to paint a broad picture of site conditions: the types of soils and the complexity of subsurface conditions. With the knowledge obtained from the preliminary survey, a detailed ground investigation can be planned and carried out.

Laboratory testing may proceed alongside site work, but more usually it is not begun until all samples have been obtained. An appraisal of conditions encountered and the range of samples obtained can then be made before the program of laboratory testing is drawn up.

THE SITE INVESTIGATION REPORT: HOW THE GEOTECHNICAL ENGINEER INTERPRETS THE SITE INVESTIGATION RESULTS AND COMMUNICATES HIS FINDINGS

Once the testing program has been completed, the engineer is faced with a mass of results which he must interpret and use to make recommendations. For a simple site with a few trial pits or boreholes this is a relatively simple problem but with large sites and complex projects, coupled with variable ground conditions, this stage can be daunting. Soil conditions vary both with depth and across the site; there may be several soil types, whose depths and thicknesses vary from borehole to borehole; and test results may show wide variation even within a single soil type. His usual recourse is to use trial hole information to draw cross sections, showing the different soil types. Test results can be marked on these so that any trend, with depth or plan position, is highlighted. For clays, plots of strength and compressibility against depth are often useful. The problem is a three-dimensional one and isometric projections can often be used to give a more graphic impression of ground conditions.

In choosing design parameters and making recommendations, the engineer must allow for the worst likely conditions. Where the odd very poor result is obtained he must decide whether it is an anomaly to be ignored, or whether it must be taken into account. Failure is likely to take place within the weakest material so it is the poorest results which are important; but to base the design on an isolated poor test result could lead to expensive over-design. It often helps to consult drillers' site records and laboratory test sheets which may indicate whether the poor results were due to exceptional ground conditions or testing problems.

Once he has sifted through all the information, the engineer will have established a soil profile (sometimes accurate, sometimes conjectural) and will have decided on suitable design values for the relevant parameters for each soil type: strength, compressibility, permeability, and so on. In some cases, especially for compressibility, he may prefer to establish a range of values. These design parameters can now be used to calculate values which are of direct interest to the project designers: types of foundations, suitable founding depths, allowable bearing capacity, amount and rate of settlement, suitable slope angles for excavations and embankments, pumping capacity required to keep excavations drained, or whatever else is required.

All the information obtained can now be gathered together and presented in the site investigation report. Reports fall broadly into two categories. The first kind simply reports on the site and laboratory work that was carried out. It gives trial pit and borehole logs, and field and laboratory test results. It may contain a summary of ground conditions encountered but no attempt is made to interpret the findings. This is often called a 'factual' report. The second type of report is usually referred to as a 'comprehensive' or 'engineering' report and additionally includes an appraisal of the findings of the site investigation and recommendations for design and construction. A good report will also state the assumptions made in reaching the recommendations and a discussion of problems that might arise during construction, with an indication of what action should be taken if any of these problems occur.

The choice between the two types of report depends on the client. Those who have their own geotechnical engineers will usually prefer to make their own assessment. They may also want their engineers to participate in site supervision, inspect the soil samples and decide on the laboratory testing program.

USE OF THE COMPREHENSIVE SITE INVESTIGATION REPORT BY THOSE INVOLVED IN THE PLANNING, DESIGN AND CONSTRUCTION OF THE PROJECT

This seems easy. You just flip through to the "recommendations" section, read the bit about allowable bearing pressures and foundation depths, and put the report in the filing cabinet.

This common approach may prove adequate on mundane projects where relatively light structures are built on good ground using simple strip or pad foundations. However, even the most mundane, straight-forward-looking projects can run into difficulty because of lack of attention to the ground conditions. It is well worth spending time looking through the report, especially at the borehole logs, site plan, any ground cross-sections, and graphs or tables of results; never mind the writing, look at the pictures. This will give you a feeling for the ground conditions at the site; types of soil, variability, and groundwater conditions.

An important point to check is how the recommended bearing capacity has been arrived at. Is it based on ultimate bearing capacity, with a suitable factor of safety? If so, what are the resulting settlements estimated to be. Have settlements been seriously considered? If settlement is the limiting factor in determining the allowable bearing capacity, is the assumed tolerable settlement reasonable for the type of structure proposed?

Recommendations of founding depths should also be looked at with caution. How were they arrived at? Has frost susceptibility been taken into account? Are the soils highly plastic and, if so, will the soil shrink and swell in the zone of seasonal moisture variation, possibly causing cracking of structures? If this problem exists, have foundations been located below the zone of seasonal moisture content changes?

Variations in soil conditions can also cause problems due to excessive differential settlement, even though total settlements are not large. Is variability a problem on your site and, if so, is it dealt with in the recommendations?

Is there any comment on groundwater conditions? Will special precautions be needed, such as pumping or de-watering, or should construction be scheduled to begin in the early summer to overcome this?

In soft or expansive clays, ground-bearing floor slabs may move vertically relative to the walls. This is due to such factors as the difference in thickness of compressible material beneath the walls and the floors; moisture content changes in the soil beneath the floor slabs; and differences in loading. Could this be a problem on your site and has reference been made to it?

Of course, a good report will deal with all these and any other relevant points; but sadly not all reports are as good as they might be. In many cases lack of proper funding is the cause; the amount of detail in the report reflects the amount the client is prepared to pay. Lack of communication between the geotechnical engineer and the project designer and a lack of understanding of each other's problems also take their toll. This can be a particular problem when the engineering interpretation is left entirely to a specialist contractor. For this reason, many firms employ a geotechnical engineer, either in-house or on a consultancy basis, even though they are not primarily involved in geotechnical projects. This is less of a problem for large organisations with their own geotechnical sections but even here problems of communication can arise between sections or individuals.

Chapter 2

TYPES OF FOUNDATIONS AND RETAINING WALLS

SCOPE

The bulk of this chapter is devoted to a review of foundation types, their applicability and the implications of the choice of foundation type on the site investigation work. Earth retaining walls are similarly reviewed in the latter part of the chapter: although not strictly foundations, they are foundation-like structures and are needed on many civil engineering projects.

THE IMPORTANCE OF FOUNDATION CONSIDERATIONS IN THE MAJORITY OF SITE INVESTIGATIONS

The purpose of a ground investigation is to establish the soil and rock profile beneath a proposed structure and to determine the engineering properties of these materials in order that the effects of the structure on the ground, and of the ground on the structure, can be predicted, or at least assessed. It follows that the ground investigated must include the entire volume of soil which is likely to be affected by the project, or which might affect the project.

In terms of foundations, this is usually taken as the volume of ground beneath the foundation which is significantly stressed by the applied loadings. Typically, stresses are considered significant when vertical pressures are greater than 10% of the pressures beneath the foundations. A useful rule-of-thumb is that the depth of soil investigated should be at least one and a half times the width of the loaded area, measured from the underside of the foundation. Of course, this often has to be modified substantially in the light of site conditions: boreholes or probes should never be terminated in weak ground. At the other extreme, there is no need, with most investigations relating to foundation work, to continue drilling once sound rock has been encountered.

With isolated shallow strip or pad foundations it is easy to interpret the one-and-a-half-width rule providing some reasonable assumptions can be made about foundation depths and widths initially. As ground conditions deteriorate, foundation widths and depths become less certain and the soil depth investigated has to be increased. With shallow raft foundations interpretation of the rule is again easy but with closely-spaced pad foundations the depth of stressed ground is less easy to estimate because of interaction between the foundation loads, as shown in Figure 2.1.

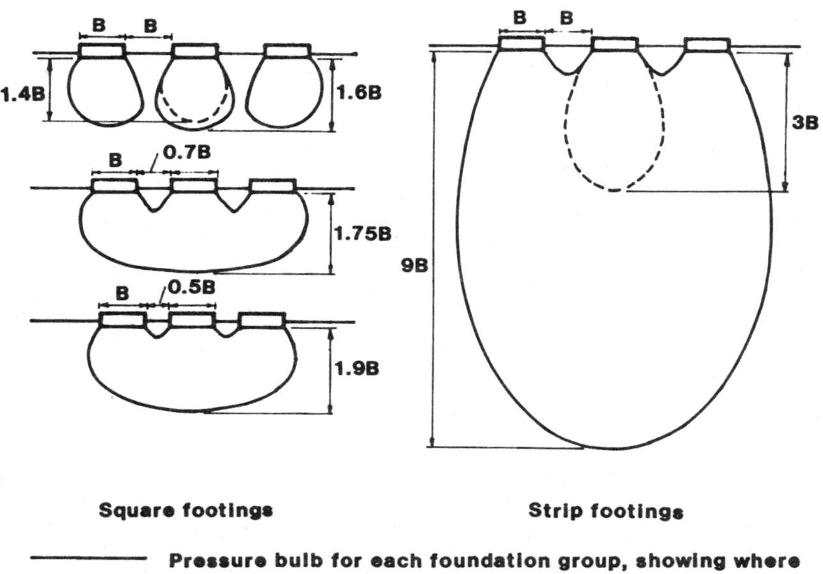

Figure 2.1 Pressure bulbs beneath foundations and foundation groups

Piles are another form of foundation where it is difficult to estimate the depth of drilling or probing needed. The complexity of the way in which load is transmitted from pile to soil, by both shaft friction and end bearing, makes the required depth of piling difficult to estimate at the site investigation stage. In addition, the cumulative effect of stresses imposed by closely-spaced piles within a group means that the soil may be significantly stressed to some depth below the base of the piles. As a result it is often impossible to make more than the roughest estimate of the total depth of soil likely to be affected by a pile group. The depth investigated should extend considerably below the maximum likely depth of the piles and well into good bearing material.

Where deep basements are required, rock head and rock quality may be important not because of its bearing capacity but because it must be excavated. Also, the weight of ground removed from a deep basement may exceed the total weight of the structure being built. This may lead to uplift problems and a requirement for ground anchors. The ground investigations will obviously have to include testing of the ground below the basement to obtain information for soil or rock anchor design.

From these examples, it can be seen that the type of foundation proposed will have a profound effect on the depth of soil investigated and, possibly, on the site investigation methods used.

REVIEW OF FOUNDATION TYPES

We have divided foundation types into two sections: those that are commonly used for domestic and light industrial buildings; and those that are used for heavy structures. The reasoning behind this is that the first category, which includes strip, pad and raft foundations and short bored piles, covers foundation types which are commonly encountered and which (rightly or wrongly) are often designed with little or no consultation with specialist geotechnical engineers. The second category, foundations for heavy structures, includes spread foundations, deep basements, piles and caissons, and specialist geotechnical knowledge is required in the design. This leads to some inconsistencies in that raft and pad foundations may be used for heavy structures, and pad footings may also be constructed as deep foundations. It also artificially separates short bored piles from piles generally. However, we feel that the advantage of categorising foundations in this way outweighs the inconsistencies for the purposes of this book.

The review also includes brief descriptions of earth retaining walls. Except for fairly low rigid or crib walls, these usually need to be designed by a specialist geotechnical engineer, but some knowledge of the types available and their applications is useful.

DOMESTIC AND LIGHT INDUSTRIAL BUILDINGS

Strip foundations

These are used beneath load-bearing walls to low and medium rise buildings made typically of brickwork or blockwork. They may also be used beneath rows of closely-spaced columns. They are the most common form of foundation and are suitable for all but the poorest ground conditions.

Examples of traditional strip foundations are shown in Figure 2.2 (a) and (b). The width of the concrete base is adjusted to give a suitable bearing pressure but a minimum width of 450mm is usually specified for practical purposes. If the foundation trench is less than this, there is insufficient room for bricklayers to work when building up the walls.

A founding depth of 0.4-0.5m is adequate for most conditions but foundations may need to be taken below the depth of frost penetration in frost susceptible soils; below seasonal moisture content changes in expansive soils; and below any upper weak layer, to reach an underlying stronger soil.

Where ground is free from deleterious substances such as sulphates, chlorides and organic material, well-compacted ordinary Portland cement concrete to a mix of 1 to 9 by volume (1 to 12 by weight) is sufficient. Ordinary quality bricks are usually satisfactory where brickwork will not be saturated for long periods and will not be exposed to frost action

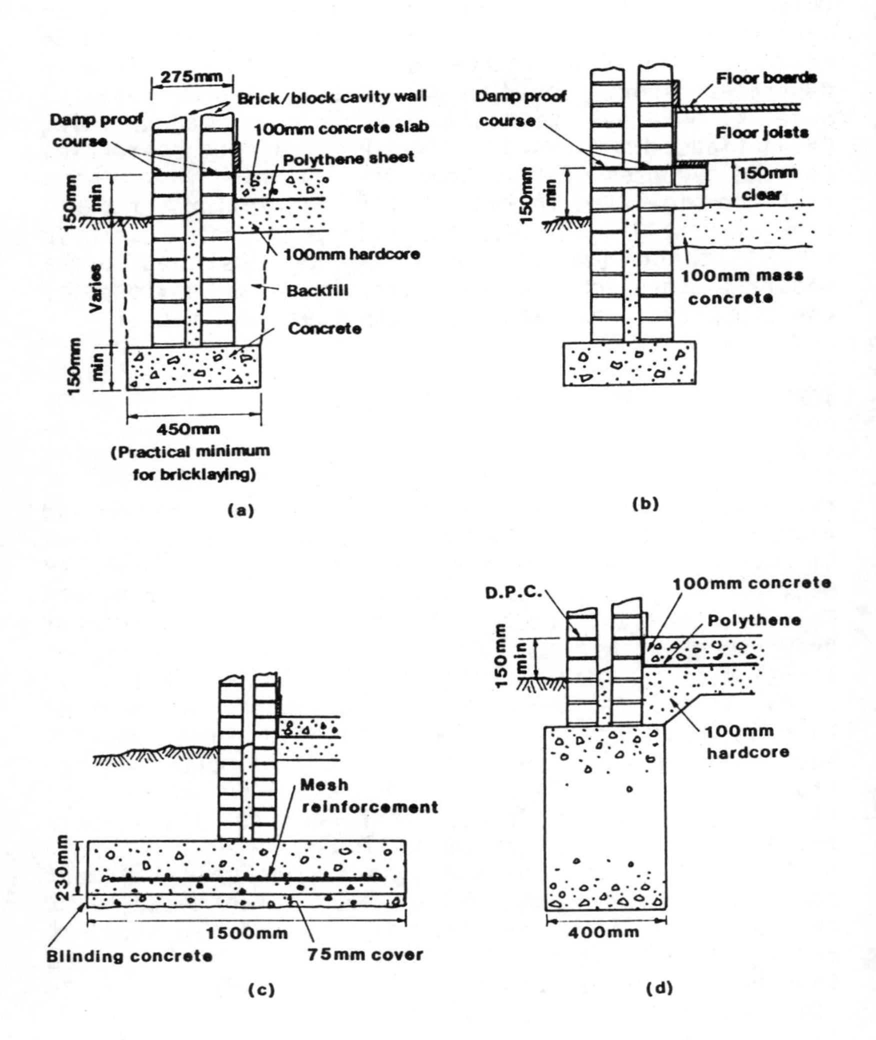

Figure 2.2 Typical strip foundation designs: (a) traditional strip with ground-bearing concrete floor; (b) traditional strip with suspended floor; (c) wide strip with reinforcement; (d) narrow strip with ground-bearing concrete floor.

when saturated. Special bricks are required in wet conditions and sulphate bearing soils.

The base of the foundation should be level and at a fairly constant depth below ground surface. This will necessitate stepping the foundation on sloping sites.

In poor ground conditions the bearing pressure can be reduced by increasing the width of the foundation and it may be necessary to include transverse reinforcement in the base of the concrete to prevent cracking, as shown in Figure 2.2 (c).

In good load-bearing soils narrow strip foundations may be used as an alternative to traditional construction. A narrow trench is dug, usually by mechanical excavator, and filled with concrete to just below the ground surface, as shown in Figure 2.2 (d). The trench width is typically 400mm: because it does not need to be wide enough to provide working space for a bricklayer, the amount of excavation is reduced and the extra concrete is cheaper than the brickwork it replaces. The method is particularly appropriate in expansive clays where foundations need to be taken down to 1m or more below ground level, or where the upper metre of soil is unsatisfactory as a foundation material. As depth increases, narrow strip foundations become increasingly more economical in stiff clays than traditional strip foundations.

Where foundations need to be deeper than about 1.5-2.0m, all forms of strip foundation become expensive and may suffer from differential lateral pressures in expansive soils. Lateral thrust can be reduced by backfilling the trench of traditional strip foundations with loose sand or, with narrow strip foundations, by placing blocks of expanded polystyrene down the sides of the trench. Where variable ground conditions occur, such as soft patches of soil or abrupt changes in bearing stratum, the concrete strip should be reinforced with steel bars or mesh.

Pad foundations

Pad foundations are used as an alternative to strip foundations for framed structures, which transmit loading to the foundation through individual columns. Small pad foundations carrying light loads are built of mass concrete. In order to keep tensile stresses on the underside of the base to an acceptable minimum, the thickness of the foundation is designed on the assumption that load is transmitted at 45^0, as indicated in Figure 2.3. As loads and footing size increase, thicknesses designed on this basis become excessive, so steel reinforcement is used to take the tension on the bottom face due to bending stresses. Shear reinforcement may also be needed to avoid punching failure.

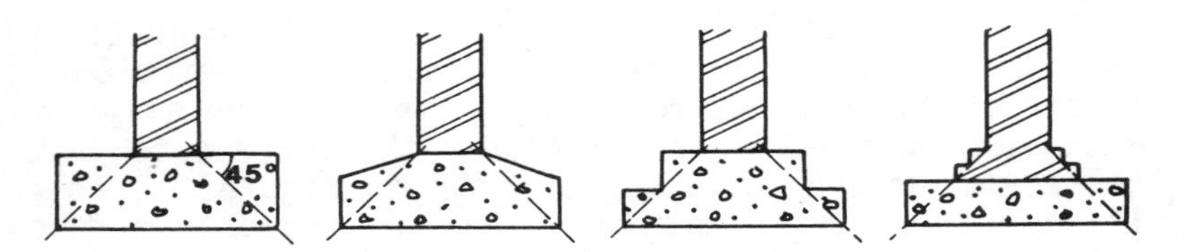

Figure 2.3 Examples of foundation thicknesses, designed on the assumption that loads are spread at 45^0*, as indicated.*

Raft foundations

On ground with very low bearing capacity or where excessive variations in ground conditions would lead to unacceptable differential settlements, raft foundations are used as an alternative to strip or pad foundations. It is sometimes suggested that raft foundations should be used where individual strip or pad foundations would occupy more than 50% of the floor area of a building. However, raft foundations are more than just a joining-together of ordinary shallow foundations: they provide much greater stiffness and consequently need considerably more steel reinforcement.

The simplest form of raft is a plain slab, reinforced top and bottom, which also forms the floor slab of the building. This is suitable for light structures where the problem is one of low bearing pressures rather than large differential settlements. The base of the slab is very shallow, which may be a problem where the upper layer of soil is weak, or in frost susceptible soils. Also, the slab projects beyond the building above ground and this may be unacceptable from an aesthetic point of view. On the other hand, very shallow foundations are easier to build where there is a high water table and they help minimise structural damage due to mining subsidence or earthquakes.

To overcome the objections of the very shallow raft, it may be located deeper with ground floors placed on a layer of compacted fill as shown in Figure 2.4 (a). Alternatively, suspended wooden floors may be used. In frost-susceptible soils it may be necessary to place the edges of the raft on a layer of granular fill or lean concrete, as indicated in the figure.

By stepping down the outside of the raft, as shown in Figure 2.4 (b), the outer edge is kept below ground level whilst the ground floor is kept above the outside ground level and sealed from water penetration. A stiffening beam can be incorporated as part of the step-down feature, as shown. For heavier loads and more compressible soil conditions, much heavier, stiffer beams are used, as shown in Figures 2.4 (c) and (d).

Short bored piles

These can be used to carry loads through soft upper layers to firmer strata beneath. This allows construction over, for instance, filled areas and former ponds or boggy areas. They are particularly useful in expansive clays as a cheaper alternative to conventional strip footings where foundations need to be taken below the zone of seasonal moisture content variation.

Construction consists of simply augering a hole and filling it with concrete. A truck-mounted spiral auger is often used but for small jobs, or when access is difficult, hand augers may be used for piles up to 350mm diameter. Casing may be necessary in permeable soils below the water table. A steel reinforcement cage may be inserted in the hole but reinforcement is not required unless tension forces are expected. Tensile uplift forces may occur in expansive clays as the moisture content increases and the surface layer of clay expands. Settlement of loose layers of fill, or drying and shrinkage of expansive clays, may result in negative skin friction in the upper part of the pile, increasing the load on it. These problems can be reduced by placing a polythene sheet around the upper

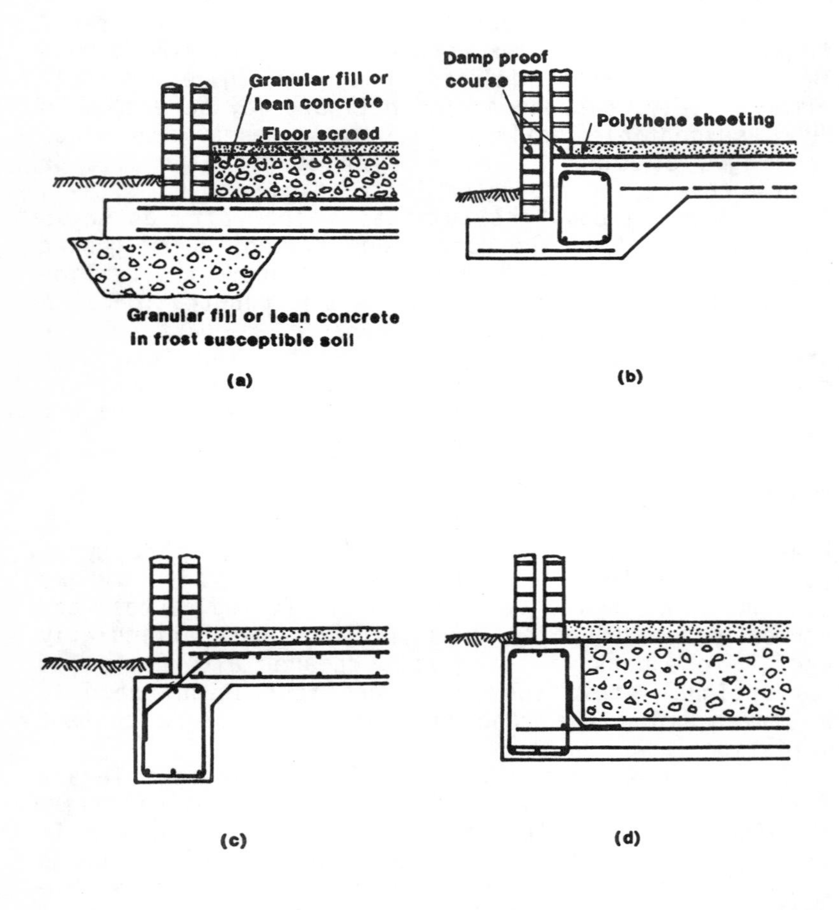

Figure 2.4 Raft foundations: (a) plain slab with raised floor; (b) stepped raft with stiffened edges; (c) raft with edge-beam for heavy loads; (d) alternative upstand edge beam construction with floor screed on granular fill or lean concrete.

section of hole before concreting. Alternatively, if a permanent casing is used it may be bitumen coated over the upper section.

Piles are connected by a capping beam as shown in Figure 2.5. Because load-bearing walls are effective at spanning between piles, the capping beam can be of lighter construction than is normally employed in pile design.

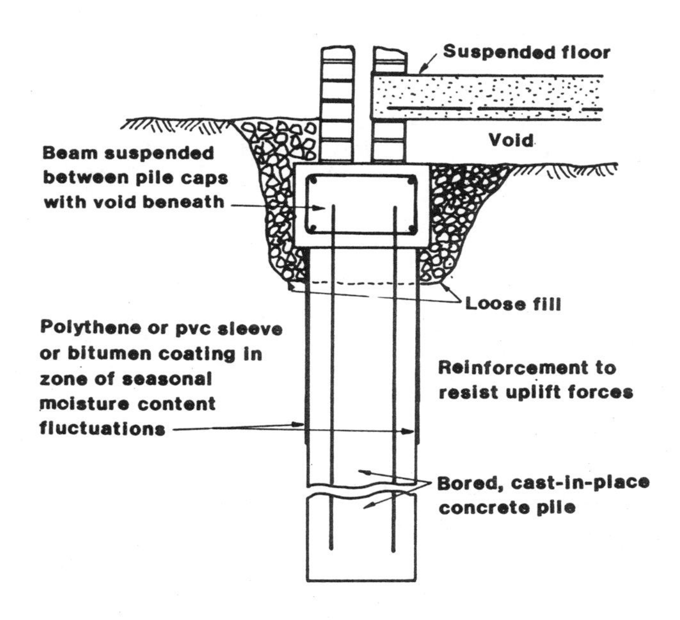

Figure 2.5 Short bored pile, showing detailing suitable for use in expansive soils.

DEEP FOUNDATIONS FOR HEAVY STRUCTURES

Spread foundations

These are used to spread column loads in the same manner as the pad foundations described earlier for light structures, and design considerations are broadly similar. However, where heavy foundation loading occurs, pad foundations may be much larger and deeper than is the case for light structures and are likely to be constructed of reinforced concrete. Where very heavy loads are to be supported on wide foundations, a steel grillage may even be used, surrounded by concrete for protection. The depth to which spread foundations may be economically constructed depends largely on the space available for excavations, allowing for the need to provide stable slopes, and on whether dewatering is required.

Deep basements

Like raft foundations, deep basements spread the foundation loading over a wide area. In addition, since soil is removed and replaced by a void, they reduce net loading on the ground. Also, because of the depth of the box formed by the foundation base and sidewalls, they provide extremely rigid foundations.

There are two main reasons for constructing deep basements: to reduce loading on the ground; and to provide extra space. Where reduced loading is the only objective, maximum rigidity can be achieved with lowest weight and least cost by using cellular construction, as indicated in Figure 2.6 (a), but a heavier, more costly, open structure may be preferred so that the space created can be utilised. Very deep basements which are built primarily to provide extra space may result in the removal of more weight of earth than the combined weight of the foundation and the structure it supports. This can cause swelling of the base and, in weak soils, there is a tendency for the structure to float upwards. To counteract this, ground or rock anchors or anchor piles may be needed to hold the foundation in place, as shown in Figure 2.6 (b).

Deep basements may be constructed in a number of ways. The simplest method is to build the basement in an open excavation. Where space is limited, bracing, cantilever sheet piling, contiguous piling or diaphragm walls may be used to support the sides of the excavation. Basements may also be constructed in the form of caissons.

Caissons

Caissons are used primarily for construction in water, typically to support jetties and wharfs.

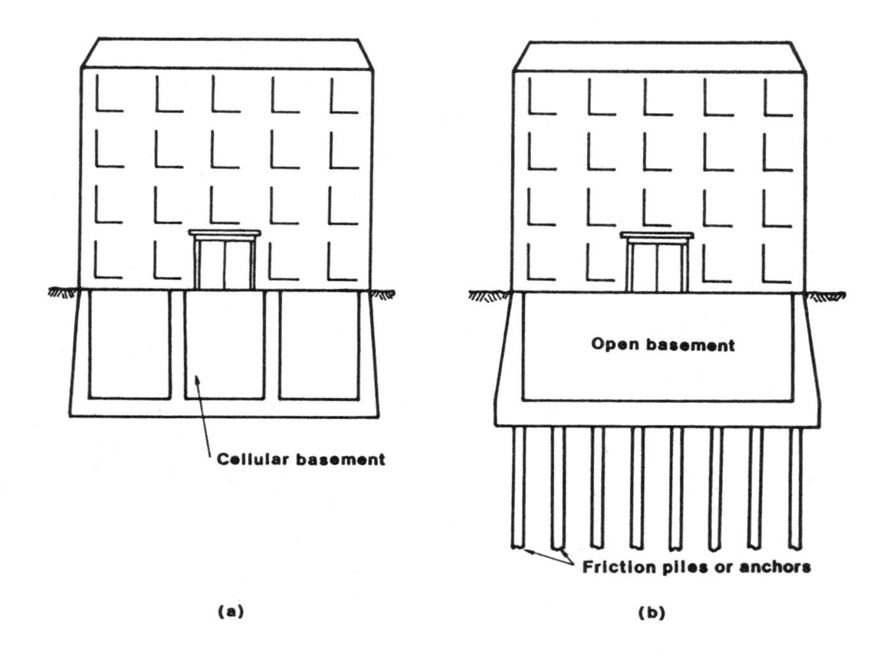

Figure 2.6 Deep basements: (a) cellular basement; (b) open basement with friction piles or anchors to resist uplift.

The distinguishing feature of caissons is that they are constructed above ground, in dry conditions, and are then lowered into place. They can be divided into three principal types; closed or box caissons, open caissons and pneumatic caissons.

Box caissons are, as the name implies, built as concrete boxes with closed bases. They are suitable as foundations in waterways where erosion of the bed is not a problem. The box is built on land, floated into position and then ballasted on to a prepared surface as shown in Figure 2.7. Depending on river bed conditions, the surface may be a levelled-off portion of bed, a layer of compacted gravel, or a concrete plinth, cast under water, which may be tied to the river bed by piles.

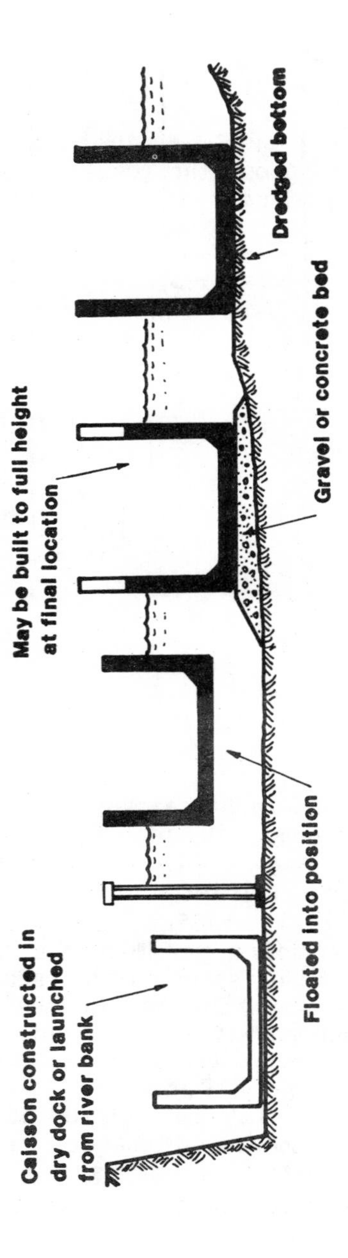

Figure 2.7 Box caissons: construction and placement.

Open caissons, which may be used on land or in a waterway, are precast or cast insitu and sunk by excavating material from the centre, as shown in Figure 2.8. When the caisson reaches the required depth, a concrete plug is poured in the base to seal it and prevent the caisson sinking further. Excavation may be by hand, by mechanical grab or, in sands, by washing out material and pumping out the resulting slurry. The cutting shoe and lower part of a caisson is usually constructed using inner and outer skins of steel plate with the space between filled with concrete. Upper walls are typically cast-in-place reinforced concrete.

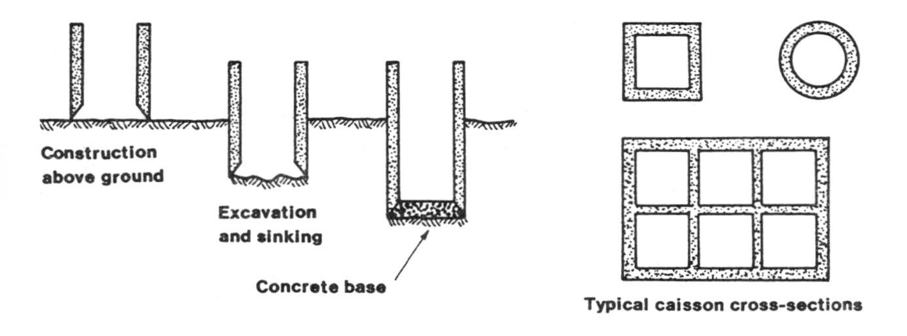

Figure 2.8 Open caissons: sinking procedure.

Pneumatic caissons have a sealed working area at the base so that excavation can be carried out under compressed air. This prevents or reduces seepage flows into the excavation, giving drier working conditions in silts, sands and gravels. Since seepage forces can cause flow of these soils up into the base of the excavation, pneumatic caissons also reduce loss of ground and consequent settlement in the surrounding area. All men and materials, including the spoil produced, must pass through air locks. This slows down working and increases costs.

Piled foundations

Pile design and construction is a vast subject about

which many books have been written. Piles are usually designed and installed by specialist contractors but it is useful when planning jobs which may involve piling to know the types of piles that are available and their relative merits. Piles can be divided broadly into two types; driven piles and bored piles. Within these two categories is a bewildering variety of types, produced by numerous manufacturers and piling contractors to cover many different soil conditions and uses, and such problems as site access and the effects on surrounding structures. A brief review of the more common pile types is given in Table 2.1.

Once installed, piles act primarily in one or both of two ways; as end-bearing piles or as friction piles. Where the base of the pile rests on rock head or in a dense sand or gravel, the load is transmitted through to this stratum with little or no bearing capacity from skin friction, as illustrated in Figure 2.9 (a) and (b). Such piles are termed end-bearing piles. Piles driven into clay soils derive the majority of their bearing capacity from skin friction or adhesion and relatively little from end-bearing, as illustrated in Figure 2.9 (c) and (d). These are termed friction (or floating) piles.

Piles have a number of applications and are used in numerous situations and ground conditions; hence the wide variety of piles on offer. Primary reasons for installing piles are: to increase bearing capacity; to decrease settlements; to provide lateral resistance; and to resist movement of the foundation due to moisture content changes in expansive soils. Driven piles have also been used to densify soils, particularly loose sands and silts. However, with the development of alternative methods of treating ground, such as the use of stone columns, deep vibratory compaction, and deep compaction by falling weights, the use of piles for ground compaction is no longer competitive.

Examples of soil conditions in which piles might be used are illustrated in Figure 2.9. Settlement of the upper layer of soil after the pile has been installed might occur where the upper clay layer is very soft, (examples (a), (b) and (c)) or where the pile has been driven through fill (example (e)). This causes an extra downward drag on the pile, known as "negative

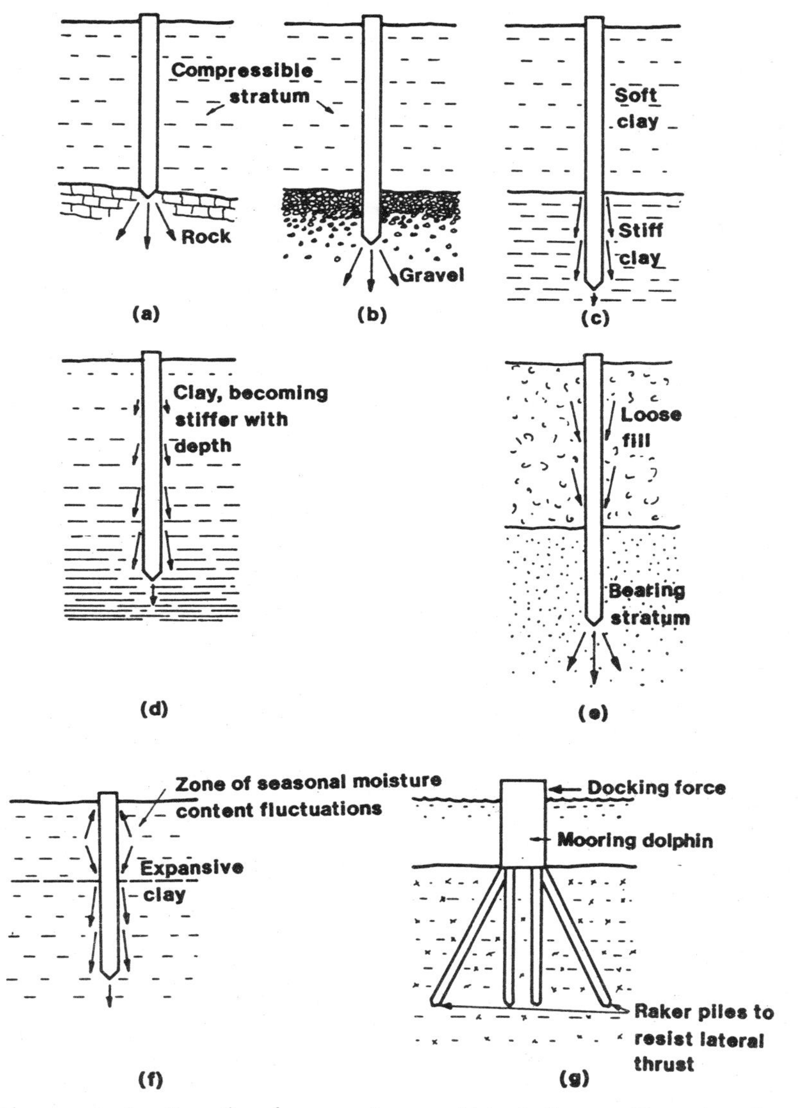

Figure 2.9 How loads are transmitted from piles to the surrounding ground: (a) and (b) end-bearing piles which terminate in rock or granular soil; (c) and (d) friction piles which terminate in clay; (e) and (f) the effects of loose or expansive soils; (g) raker piles to resist lateral thrust.

TABLE 2.1
BROAD CLASSIFICATION OF PILE TYPES

DRIVEN PILES	
PREFORMED	**CAST IN PLACE PILES**
Steel: may be H-section, box or tube, driven 10-30m. Easy to splice and cut. Working stress typically 80MN/m^2, giving a high working load of 40-120t. Soil displacement is small. Particularly suitable where hard driving is required but may be damaged by large boulders. Vulnerable to corrosion; allowable bearing capacity may need to be reduced in corrosive locations. **Concrete:** may be pre-cast or prestressed. Precast driven 10-15m, prestressed driven 20-30m. May be designed for a wide range of loads, with very high carrying capacity if required. Piles can tolerate hard driving and may be corrosion resistant but are costly; prestressed piles are difficult to splice. Piles may be damaged by handling. **Timber:** typically pine or Douglas fir, driven up to 10-20m but timber is difficult to splice. Maximum working stress about 8MN/m^2, depending on timber type. Typical load capacity 10-50t. Timber piles are comparatively low cost and submerged treated timber is resistant to decay, but above the water table, and particularly where the pile is intermittently submerged, piles may suffer attack.	**Thin steel shells:** piles are driven by means of a mandrel; a steel tube which transmits the force of the hammer directly to the pile shoe, leaving the walls relatively unstressed. After driving, the mandrel is removed, a steel reinforcement cage inserted and the pile concreted. This type is usually cheaper than thick steel shell piles but the thin shells are vulnerable to damage during driving. Piles may be used in granular soils, with a taper section, for medium loads. **Thick steel shells:** piles are driven without a mandrel, then concreted. The strength of the shell is taken into consideration when calculating the pile load capacity. Typical lengths 10-25m. Vulnerable to corrosion. **Concrete shells:** the shell is made up of short, precast, reinforced concrete segments. The pile is driven with a mandrel, as for thin steel shell piles. May be made corrosion resistant. **Withdrawable steel drive tube:** a tube is driven with the end closed by a concrete plug or detachable shoe. Concrete is rammed into the pile as the tube is slowly withdrawn. The concrete at the base can be hammered down to form a bulb, increasing bearing capacity. In weak soils, the hammering causes compaction of the surrounding soil, increasing its strength. In peats or very soft soils, or where work is carried out carelessly, soil can squeeze inwards as the drive tube is withdrawn, reducing the pile diameter. This is known as 'necking' or 'waisting'. Typical working loads range from 35t for 350mm diameter piles to 140t for 600mm diameter piles.

TABLE 2.1
(CONTINUED)

BORED PILES	GENERAL COMMENTS (ALL PILE TYPES)
Hand augered piles: holes are bored to the required depth, using a hand auger, and filled with concrete. Reinforcement is not usually used unless required to resist bending or uplift forces but dowel bars are cast into the tops of the piles, to tie in with the foundation beams. Piles are less than 5m depth, 350mm maximum diameter and are used beneath houses and low-rise, light industrial buildings. Because no casing is used and simple augers cannot easily penetrate stoney soil, piles are usually restricted in use to clays and clayey silts or sands above the water table. **Mechanically augered piles:** a variety of augers (spiral, flight, bucket) are used, attached to rotary drilling machines which are usually mounted on cranes or trucks. Piles may be used for depths up to 45m and shaft diameters up to about 2m. Under-reaming tools allow bases to be enlarged to about 7m. Augered piles are most suited to clay soils: augers cannot penetrate soil containing appreciable numbers of cobbles and boulders or other obstructions. Bentonite or casing may be used to support the sides. With flight augers, the auger itself may be used to support the sides, the auger being slowly withdrawn as concreting proceeds. **Percussion bored piles:** if ground conditions are unsuitable for hand or mechanical augering (e.g. very soft silts or clays, water-bearing granular soils) the hole can be sunk using a cable percussion boring rig similar to the shell and auger rig employed for site investigation boreholes.	**Composite piles:** Composite piles are formed from combinations of the types described above. For instance, steel or wood may be used below the water table, where it is less likely to be susceptible to attack, and concrete used for the upper part of the pile. **Choice of pile type:** The choice of pile will depend on factors such as soil type, location, accessibility, type and size of structure, and cost. In general, driven piles tend to be used in gravels, sands and silts and bored piles tend to be used in clays, particularly for heavy structures. One consideration is that with certain pile types (hollow shells, bored piles) the state of the hole or driven shell can be examined, but with other types (H-section, solid precast concrete) the engineer must trust that the pile is installed satisfactorily. **Working stresses:** The following can be used as a rough guide to the working stresses to be used for piling materials. Timber: about 8MN/m^2 Steel: about 60-80MN/m^2 Concrete: 15% of 28-day strength to a maximum of about 5MN/m^2 for precast piles; 20% of 28-day strength to a maximum of about 7MN/m^2 above prestress for prestressed piles; 25% of 28-day strength up to a maximum of about 7MN/m^2 for cast in place piles.

friction", which increases the loading on it. This can also happen in the expansive clay (example (f)) but here expansion can also have the opposite effect; as the clay expands upwards it has a tendency to lift the pile out of the ground. In all these cases the piles must be designed to resist the extra forces on them which result from ground movements, in addition to the normal structural loading. To reduce such forces the upper sections of piles may be coated in bitumen or surrounded by polythene to reduce friction in the affected lengths, although there is some dispute about the efficacy of such measures.

Choice of type depends on a number of factors such as soil conditions, loads to be carried, durability and, of course, cost. Other factors which may have to be taken into account include accessibility, the level of noise which can be tolerated at the site and the proximity of surrounding buildings, which may limit the vibration or ground heave that can be accepted.

Driven piles are used in loose silts, sands and gravels and in soft to firm clays. They are particularly suitable in waterlogged ground and in construction over water where driving equipment must be located on platforms and it is useful to have a length of pile projecting out of the ground. Where piles have to be driven to great depths, steel H-piles are often used. Driven piles are not suitable in ground containing boulders or where ground heave is a problem. Noise of driving may also limit their use. Another disadvantage is that it is difficult to vary the lengths of the piles to suit different driving lengths. This is overcome in some types of driven cast-in-place piles that use withdrawable steel tube or are made up of short precast concrete shells.

Because driven piles are usually driven to a "set" - a certain number of blows per centimetre at the cessation of driving - the engineer can usually feel some confidence in the pile performance. However, what has actually happened to the pile in the ground is anybody's guess, particularly where solid piles are used. Where hollow shells are used, the shell can be inspected after driving by lowering a lamp or camera down it. Where piles are cast in place using a withdrawable steel tube, there is always the worry that "necking" will take place in soft soils. This can now be checked using a meter which detects vibrations in the pile when the pile head is struck with a hammer.

Bored piles are suitable in stiff clays and for soils containing boulders which would stop or deflect driven piles. They can be formed to large diameters and great depths, so are suitable for very heavy loads. In common with some of the driven, cast-in-place piles, necking can be a problem in soft clays. There is not the assurance of their having been driven to a set as with driven piles but the excavated shafts and the spoil from them can be examined. Headroom and site access requirements for installation are much less than for driven piles, and they are much quieter to install, with no ground heave.

EARTH RETAINING WALLS

Probably the most common type of retaining wall is the rigid wall, typically made of mass or reinforced concrete, which depends for its stability on having a broad base and on the weight of the wall itself. Crib walls are a variation of the rigid wall, in which the concrete structure is replaced by an open framework and the weight of the soil contained within the frame replaces the weight of the concrete structure. Examples of rigid walls and a crib wall are shown in Figure 2.10 (a) and (b). These types of walls are typically constructed up to 6m high, though they may be higher.

An alternative form of construction for walls up to about 5m high is the cantilever sheet pile wall, illustrated in Figure 2.10 (c). Sheet pile walls have the advantage that they can be driven into the ground first and then material can be excavated from the front, so that support to material behind the wall is not lost at any stage of construction. As wall heights increase, bending moments on cantilever walls become too high, making them uneconomic or completely impractical. To overcome this, the top of the wall is tied back to an anchorage, as shown in Figure 2.10 (d). This is often referred to as an anchored bulkhead and allows walls to be built up to about 10m. Beyond this height, multiple layers of anchors are usually needed. As an alternative to the anchor block and tie rods

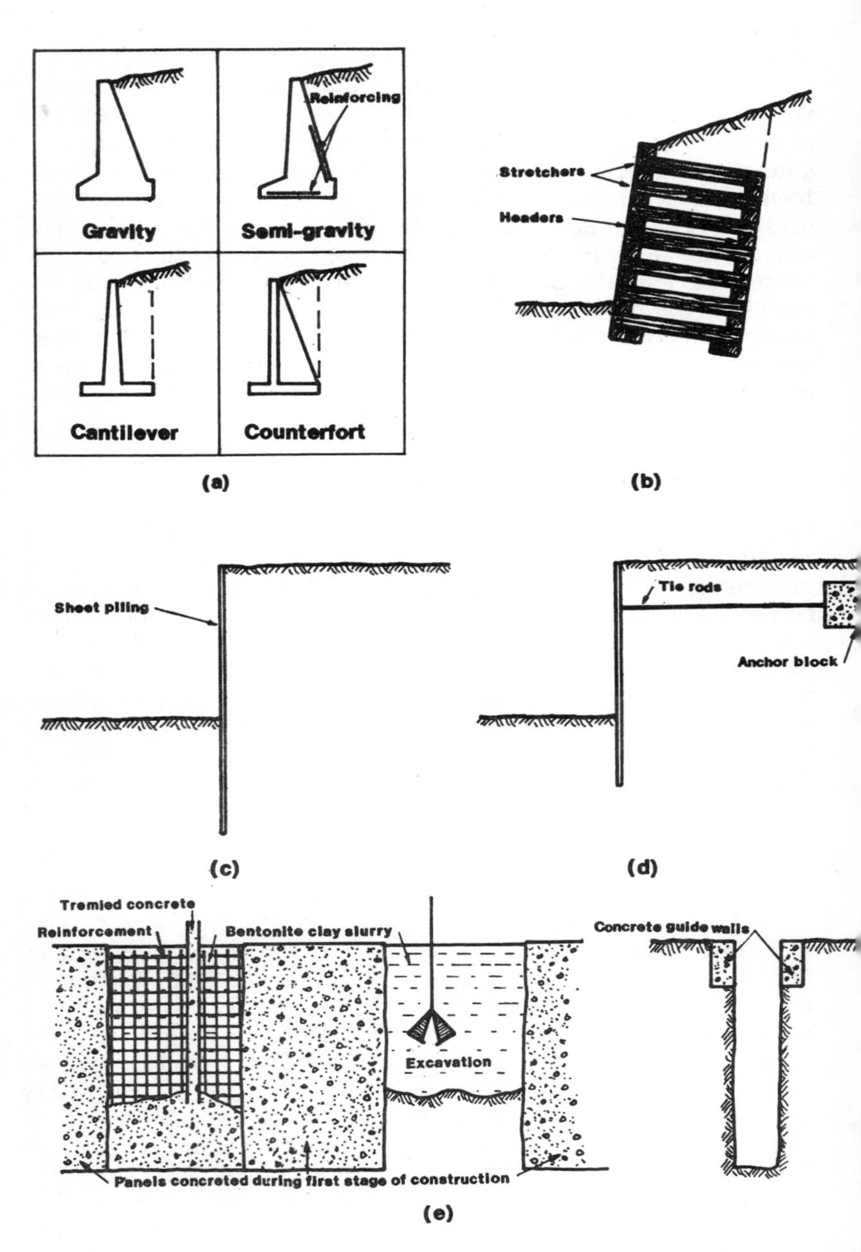

Figure 2.10 Earth retaining walls: (a) rigid walls (b) crib wall; (c) cantilever wall; (d) anchore bulkhead; (e) diaphragm wall.

illustrated, ties may be in the form of rock bolts, ground anchors or inclined piles.

A method of forming retaining walls for deep basements is the diaphragm wall. The principal feature of the diaphragm wall is that a trench for the wall is excavated using a bentonite clay slurry to support the trench sides. Excavation is carried out in alternate bays and, as shown in Figure 2.10 (e), once a bay has been excavated, steel reinforcement is placed into the bentonite-filled trench and concrete is placed by tremie pipe so that it displaces the bentonite upwards. Diaphragm walls can be installed where obstructions prevent driven sheet piling from being used. They produce less noise and vibration during installation and, being more rigid, cause less settlement to the surroundings; an important consideration in city areas. A disadvantage is that guide walls must first be built, as shown in the figure, and these can be very costly. Ground anchors can be used to tie back high walls.

Another alternative for deep basements is to use a contiguous bored pile wall. This consists of a row of piles which are touching or almost touching each other. Initially every second pile is installed, with the intermediate ones subsequently added. A slight variation of this is the secant pile method, in which the initial piles are at slightly closer spacing so that part of the pile shafts have to be chiselled away to install the intermediate piles. This increases costs but makes the wall more watertight.

Chapter 3

THE SITE INVESTIGATION

SPECIFYING WHAT IS REQUIRED FROM THE INVESTIGATION

A site investigation is undertaken in response to a need; the need to assess the suitability of a site for a particular project. When specifying how work is to be phased, it is useful to keep in mind the various phases of the site investigation process and how these fit into the overall planning of the project. A typical site investigation process is shown in Figure 3.1, which indicates the type of information obtained from each phase. Though typical, this sequence will not be followed for all site investigations. For instance, a preliminary study may be required to determine whether a particular project is feasible before detailed plans are made. On other projects the reconnaissance may be carried out before the desk study, or the ground investigation may be carried out as one phase instead of two.

The first thing to consider is what information is required by the designer. It is also important to consider how sensitive the structure will be to the ground conditions and soil properties. Obviously, more sensitive designs need more precise soils information. The next stage is to obtain as much information about the site as possible from existing records. These will include geological and topographical maps, geological memoirs, previous site investigations in the area, local authority or private records, and anything else which is available. The amount and type of information available will vary greatly from site to site but it should at least give an indication of likely ground conditions. A site visit will then give added information on the topography, vegetation, and general conditions at the site. A simple walk around a site and its surroundings can provide many important details and gives a "feel" for the problem which no amount of reporting can convey. The experienced eye can pick up many clues about subsoil conditions and where problem areas may occur.

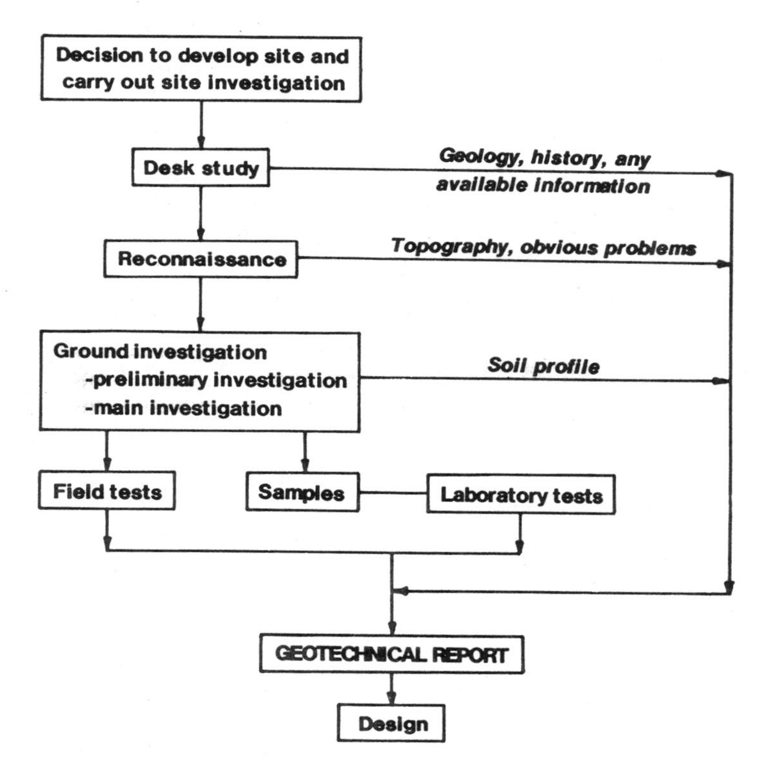

Figure 3.1 Typical organisation for a site investigation

The information about the design, obtained from the desk study and gathered from the site visit, can now be combined to make an estimate of the probable ground conditions underlying the site and to assess the probable type or types of foundations that will be required. An assessment can then be made of the ground investigation methods that would be most appropriate and of the extent of investigation work required, as indicated in Figure 3.2. Once the overall geotechnical problems have been defined, the material properties (such as soil strength and compressibility) which are needed to produce a design can be defined. The tests required, and hence the type of samples needed to obtain these properties, depend on the properties themselves and on the type of soil. In their turn, the sample types needed, the soil conditions, and the depth that must be tested will suggest the most appropriate methods of exploration and the extent of work required: the type and numbers of trial holes, samples, probes, and so on.

By assembling the information described above and making some assumptions, the engineer can draw up a specification for the work required in the ground investigation. Precise numbers of boreholes, samples and tests cannot be specified but a reasonable assessment of quantities should be possible on simple projects. As projects or ground conditions become more complex, it becomes more difficult to make a reasonable assessment. To overcome this, a preliminary ground investigation is first called for in order that the main investigation can be planned with more precision. Even so, detailed decisions, such as the location of trial holes and the depths at which samples are taken must be made on site in the light of conditions encountered and cannot normally be specified beforehand.

THE DESK STUDY

The cost of obtaining information from a desk study is usually relatively low when compared with the cost of boring and sampling, and the information obtained can often increase the effectiveness of the ground investigation. It is therefore worthwhile spending some time on this phase of the work. The desk study will also provide information on such things as past use, potential geological problems, liability to flooding and mining subsidence which may vitally affect the proposed project. Table 3.1 gives a list of information which should be considered at the desk study phase. Additional considerations are needed at

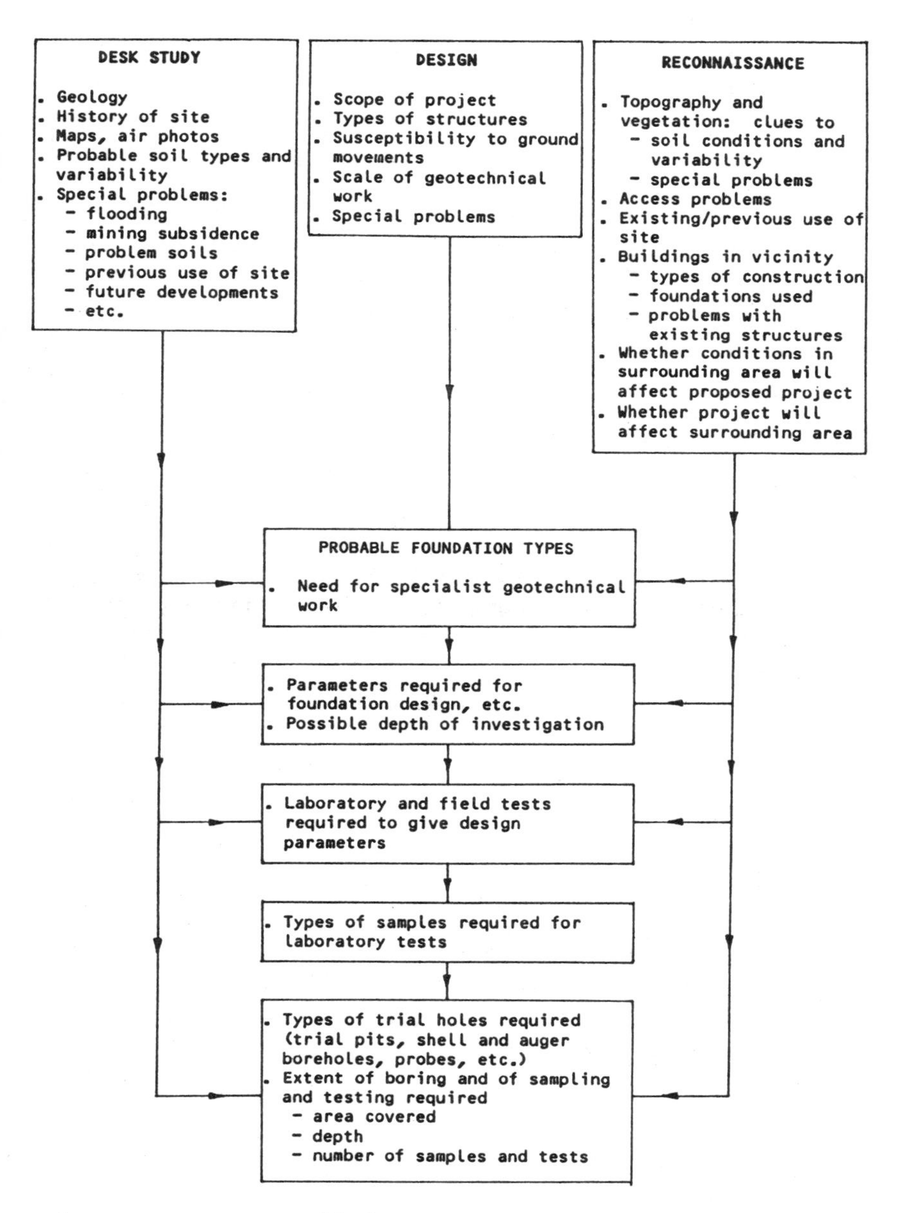

Figure 3.2 How preliminary information is used to plan the fieldwork for a site investigation.

TABLE 3.1
INFORMATION THAT SHOULD BE CONSIDERED AT THE DESK STUDY STAGE (AFTER BS 5930:1981)

A.1 General land survey
- (a) Location of site on published maps and charts.
- (b) Air survey, where appropriate.
- (c) Site boundaries, outlines of structures and building lines.
- (d) Ground contours and natural drainage features.
- (e) Above ground obstructions to view and flying, for example, transmission lines.
- (f) Indications of obstructions below ground.
- (g) Records of differences and omissions in relation to published maps.
- (h) Position of survey stations and bench marks (the latter with reduced levels).
- (i) Meteorological information.

A.2 Permitted use and restrictions
- (a) Planning and statutory restrictions applying to the particular areas under the Town and Country Planning Acts administered by appropriate Planning Authorities.
- (b) Local authority regulations on planning restrictions, listed buildings and building bye-laws.
- (c) Board of Trade regulations governing issue of industrial development certificates.
- (d) Rights of light, support and way including any easements.
- (e) Tunnels; mine workings, abandoned, active and proposed; mineral rights.
- (f) Ancient monuments; burial grounds, etc.

A.3 Approaches and access (including temporary access for construction purposes)
- (a) Road (check ownership).
- (b) Railway (check for closures).
- (c) By water.
- (d) By air.

A.4 Ground conditions
- (a) Geological maps.
- (b) Geological memoirs.
- (c) Flooding, erosion, landslide and subsidence history.
- (d) Data held by central and local authorities.
- (e) Construction and investigation records of adjacent sites.
- (f) Seismicity.

A.5 Sources of material for construction
- (a) Natural materials.
- (b) Tips and waste materials.
- (c) Imported materials.

TABLE 3.1
(CONTINUED)

A.6 Drainage and sewerage
- (a) Names of sewerage, land drainage and other authorities concerned and by-laws.
- (b) Location and level of existing systems (including fields, drains and ditches), showing sizes of pipes, and whether foul, storm-water or combined.
- (c) Existing flow quantities and capacity for additional flow.
- (d) Liability to surcharging.
- (e) Charges for drainage facilities.
- (f) Neighbouring streams capable of taking sewage or trade effluents provided they are purified to the required standard.
- (g) Disposal of solid waste.
- (h) Flood risk: (1) to proposed works.
  (2) caused by proposed works.

A.7 Water supply
- (a) Names of authorities concerned and bye-laws.
- (b) Location, sizes and depths of mains.
- (c) Pressure characteristics of mains.
- (d) Water analysis.
- (e) Availability of water for additional requirements.
- (f) Storage requirements.
- (g) Water source for fire fighting.
- (h) Charges for connections and water.
- (i) Possible additional sources of water.
- (j) Water rights and responsibilities under the Water Resources Act 1963, which controls permissions to take water from any natural source.

A.8 Electricity supply
- (a) Names of supply authorities concerned and regulations.
- (b) Location, capacity and depth of mains.
- (c) The voltage, phases and frequency.
- (d) Capacity to supply additional requirements.
- (e) Transformer requirements.
- (f) Charges for installation and current.

A.9 Gas supply
- (a) Names of supply authorities concerned and regulations.
- (b) Location, sizes and depths of mains.
- (c) Type of gas, thermal quality and pressure.
- (d) Capacity to supply additional requirements.
- (e) Charges for installation and gas.

A.10 Telephone
- (a) Address of local office.
- (b) Location of existing lines.
- (c) GPO requirements.
- (d) Charges for installation.

A.11 Heating
- (a) Availability of fuel supplies.
- (b) Planning restrictions (smokeless zones; Clean Air Act 1956 administered by local authorities).
- (c) District heating.

the desk study phase in old coal mining areas and these are discussed in Chapter 8.

As much information as possible should be obtained from existing sources. The amount of information available and the most useful sources will depend on the individual site. Some common sources of information are discussed below.

Topographical maps

The popular scale of map for most purposes is the 1:50000 but for use with site investigations a 1:25000 is more useful. Where more detail is required, maps at scales of 1:10000, 1:2500 and 1:1250 are regularly used. These show individual buildings and boundaries such as walls and fence lines. It should be remembered when using small scale maps that some features have to be shown diagramatically or in a simplified form and this can cause local distortions of scale. Large scale maps show detail more accurately but details such as field boundaries, woodland areas, buildings, and even minor roads can change within a few years so that numerous minor inaccuracies are to be expected with these maps. Out of date detail is, of course, a problem with all maps to some extent. Sequential editions of the maps should be referred to, whenever possible, as they can provide valuable information on the past use of the site.

Topographical maps have a variety of uses. Initially they can give an overall picture of the area, highlighting the inter-relation of different features in a way that may not be obvious to the more restricted view of an observer on the ground. They are also useful for planning the layout of projects and the location of the first few trial holes or probes. A study of ground contours and drainage can also give an indication of differences in ground conditions and of possible problems: soft ground in flat areas; slope stability problems on steep hillsides. This can be checked out during a visit to site and the map can be used in conjunction with observations of site conditions to sketch in possible boundaries between different soil types; a simple form of terrain mapping.

Geological maps

Geological maps are available for all areas of Britain at a scale of 1:63360 (1 inch to a mile), with some of the more recent revisions at 1:50000. The geological formations at the surface are shown as overprinting on a faintly-printed Ordnance Survey underlay map. "Drift" maps include geologically recent deposits such as river sediments, whereas "solid" maps show only the underlying main geological deposits. For most site investigations, the "drift" maps are more useful. Geological Memoirs are available with some maps. They describe the geology of the area in some detail. There are also "Handbooks of British Regional Geology" which cover all of England, Scotland and Wales as shown in Figure 3.3. More detailed maps are available for most Coal Measures areas and for some large urban areas at scales of 1:10560 (6 inches to a mile) or 1:10000. Geological maps are produced in most overseas countries but the amount of detail given and the accuracy of the information are very variable.

From the point of view of the geotechnical engineer, the information given in geological maps and reports is too general to allow him to make any sound assessment of foundation conditions. Nevertheless, it is useful to give a general indication, especially of how variable conditions are likely to be and where variations might occur. Maps also show the angle and direction of dip of sedimentary rocks; factors which affect the stability of rock excavations. It should be remembered that considerable variations of soil or rock type can occur within a particular geological formation and that the boundaries between formations may not have been accurately marked on the maps because of the inherent difficulties of geological mapping. Despite these shortcomings, geological maps are a valuable source of information which is readily available and they should be consulted on even the smallest project.

In Britain, geological maps and memoirs are available from the British Geological Survey in London, Leeds, Exeter, Edinburgh and Belfast.

Other mapping

Pedological maps, classifying surface soils in terms of

agricultural use are available for some areas. Whilst the information is not of direct interest to the engineer, variations in surface soils are related to variations in the underlying ground conditions. Pedological maps can therefore be useful when planning a ground investigation.

Figure 3.3 Areas covered by the "Handbooks of the Geology of Great Britain".

Many local authorities produce land use maps, showing areas used for housing, industry, arable farmland, grazing land, marshland, and so forth. Maps of this kind can sometime provide useful clues to ground conditions. Mineral resources maps may also provide useful clues, if available, as well as information of more direct interest, such as possible sources of construction materials. Maps of mine workings, such as those held by British Coal in Britain, are invaluable in assessing potential subsidence problems.

Aerial photographs

Photographs may be taken either obliquely or vertically, from an aircraft, but for most surveying and engineering work vertical photographs are the most common. The usual size of negative is 230mm (9 inches) square and prints are made by contact printing to minimise distortion so the scale of the photograph depends on the height of the aircraft above the ground and the focal length of the camera lens. Black and white photographs are the most common but colour photographs are sometimes used. Infra-red photography is occasionally used to pick out special features. A special film, sensitive to infra-red radiation, is used so that warm objects and healthy vegetation appear red; green and grey inanimate objects look blue; red objects such as tropical red soils look green; and cold, blue objects such as water look black. A fairly recent development has been the use of four cameras simultaneously, each fitted with a different film/filter combination so that each records an image, on a black and white negative, of a specific wave band of light. These images can be combined using enlargers fitted with colour filters to produce a "colour composite". The process of assigning certain colours to specific wavebands is known as "false colour" imagery. By carefully selecting different colour combinations, specific features can be made to show up more prominently. This technique, known as "colour enhancement" can be used to pick out such features as wet marshy areas and gravel deposits.

The use of aerial photographs to make accurate maps (photogrammetry) and the interpretation of colour composite pictures both need specialist knowledge and are best left to specialist organisations. However, simple black and white or colour photographs have many uses and a great deal of information can be gleaned from them after only a little experience. At a scale of about 1:10000, they serve as good up-to-date maps which show details down to the size of a cow, although cows, of course, cannot be relied upon to stay in the spot where they were photographed. They give a valuable overview of an area and show up differences in surface texture caused by variations in vegetation and soil conditions. This can be used, in conjunction with a site visit, to help delineate variations in subsoil conditions.

Photographs are usually taken with 60% overlap so they can be viewed through a stereoscope to give a three-dimensional view. This is helpful for such things as route planning of roads but is less useful on small sites. Photographs are available for the British Isles from organisations such as the Ordnance Survey and British Coal, and through specialist surveying companies. There is photographic coverage of substantial parts of the World but for large jobs it may be necessary, or more convenient, to commission a set specially. Where an area has been photographed at intervals over many years, a history of land use and development is available. This may be helpful in locating possible hazards such as old river channels, backfilled ponds or quarries, and buried foundations.

Other information

Site investigation reports covering nearby areas (or even the site under investigation) may be held by bodies such as local authorities, governmental departments and public or private companies. Architects, consulting engineers and contractors are especially likely to hold this type of information. Details of sub-soil conditions encountered during previous excavation work or trenching operations can be of particular value. Useful background information can sometimes be obtained from local inhabitants. Access to this knowledge will simplify the planning of the investigation, increase its effectiveness and decrease the extent of site work needed. The main problem lies in gaining access to, or even discovering the existence of, such information.

It is essential though, before beginning the ground investigation, that the presence of any buried services (gas, electricity, telephone, water or sewage) be checked with the appropriate utilities.

APPRAISING THE SITE AND PLANNING THE GROUND INVESTIGATION

Site reconnaissance

After or during the course of the desk study the engineer should visit the site and its surrounding areas: this is the reconnaissance phase. Ideally, he should have already completed the desk study but at the least he should have a topographical map, a geological map and, possibly, aerial photographs. He should walk around the site to get a general "feel" for the conditions, noting such things as landforms, drainage, vegetation and existing land use.

Probably what characterises the site most will be the topography. Is this an area of flat, open country, rolling hills, or steep slopes? How does the site area fit in with the surrounding areas? The steepness of slopes and changes in slopes are very important. Linked with topography will be drainage. Are there any watercourses on or near the site? Are there any lakes or ponds, or any water-logged areas? Surface soils give a clue to subsoil conditions and drainage. Are the soils sandy, or heavy clays, or strong stoney soils? Do soils vary from one part of the site to another? The soil conditions and drainage are reflected in the vegetation, be it natural or crops (which can include grass). Lush vegetation with large succulent leaves indicates wet conditions; gorse, thorn bushes or whispy grass tend to indicate dry conditions or perhaps disturbed ground. It is not necessary to be a botanist to pick up important clues from the vegetation and the variations from one place to another. Rock outcrops are obviously important since they indicate the type of rock underlying the site (this should be checked against the geological map), the quality of rock (how weathered) and the depth to rock head. Rock outcrops may occur as mounds or slabs of rock protruding out of the ground surface or river beds, or exposed in quarries or cuttings. Scattered outcrops across the landscape may indicate shallow, and possibly highly variable, depth to bedrock. Old quarries and cuttings give a good indication of subsoil types as well as rock quality and depth. Existing land use may affect both the ground investigation work and the proposed project. For instance, at the site investigation stage access to fields under crops may be restricted; and construction may be affected by foundations to existing or former buildings.

The various aspects discussed above are closely interrelated. This interrelation can be used to help the engineer make an assessment of subsoil conditions, including the types of problems that are likely to arise. For instance, soft, compressible soils may be a problem in flat terrain whereas slope stability may be a worry on steeper ground. Landslips are not always obvious but may be detected by such features as ground fissures, unusually steep slopes, trees at an angle on the slipped soil, and bulges of hummocky ground with poor drainage and a line of springs near the toe. Any buildings on the site should be inspected to see what kind of construction has been used and whether they are suffering from any kind of distress.

The engineer should not confine himself to the site itself but should look at the surrounding areas. Important clues such as distress to buildings, soil exposed in excavations and hillside instability may be apparent in the areas nearby but not at the site itself. Nor should the engineer be concerned only with subsoil conditions. Marks on bridges and debris in trees alongside streams may give clues to past heights of flooding. Simply talking to local people can often be enlightening. There may have been a quarry which was filled in, a history of landslips or flooding, or a stream which now flows undetected through a culvert.

A visit to the local library and council offices may also yield useful information. Strictly, such information comprises part of the desk study but it may be convenient to search for it during the site reconnaissance.

It is important to allow sufficient time to assimilate the features of the area and obtain a feel for the site. A photographic record should be kept, as well as a full set of notes. A compass, spade, hand auger and sample bags or jars may also be useful. Notes on site reconnaissance, given in B.S. 5930:1981, are reproduced in Table 3.2.

Specifying the numbers and types of trial holes

Having familiarised himself with the site and compared his findings with information obtained from the desk study, the engineer can now plan the ground investigation. Boreholes, trial pits and probes should be positioned so that all different ground conditions are covered and the boundaries between the various soil types located as accurately as possible. The scope of the investigation required can now be estimated, enabling a specification and bill of quantities for the ground investigation to be drawn up. For a simple site it may be possible to specify the number and position of pits or borings required and to give a reasonably accurate estimate of depths to be investigated and the number of samples to be taken. As conditions become more complex, it will only be possible to give an estimate of the number of trial holes needed; only the positions of the first few holes being specified initially. The number and positions of the remaining holes will depend on the findings of the first few. On large or complex projects, the ground investigation may be split into two phases: a preliminary phase with a limited number of pits and borings to establish a general indication of site conditions; followed by the main ground investigation, planned with the help of the information gained from the preliminary phase.

As a general guide, trial holes may be spaced 25m to 150m or more apart in uniform conditions but spacings of 10m or less may be needed to examine detailed problems in complex conditions. Examples of typical spacing requirements are given in Table 3.3 but it must be emphasised that the requirements of individual sites may vary considerably from those given.

Required depths depend mainly on subsoil conditions and on the type of proposed structure or development. Where poor foundation material is encountered, such as soft clay, loose sand or uncompacted fill, borings should be extended through this to reach sounder material. If great depths of soft, compressible or loose material are encountered, borings should be taken down to the depth where imposed stress from the proposed structure is negligible.

TABLE 3.2
NOTES ON SITE RECONNAISSANCE (AFTER BS 5930:1981)

C.1 Preparatory

(a) Whenever possible, have the following available: site plan, district maps or charts, and geological maps and aerial photographs.

(b) Ensure that permission to gain access has been obtained from both owner and occupier.

(c) Where evidence is lacking at the site or some verification is needed on a particular matter, for example, flood levels or details of changes in site levels, reference should be made to sources of local information such as: Local Authority, Engineer's and Surveyor's Offices, early records and local inhabitants.

C.2 General information

(a) Traverse whole area, preferably on foot.

(b) Set-out proposed location of work on plans, where appropriate.

(c) Observe and record differences and omissions on plans and maps; for example, boundaries, buildings, roads and transmission lines.

(d) Inspect and record details of existing structures.

(e) Observe and record obstructions; for example, transmission lines, telephone lines and ancient monuments, trees subject to preservation orders, gas and water pipes, electricity cables, sewers.

(f) Check access, including the probable effects of construction traffic and heavy construction loads on existing roads, bridges and services.

(g) Check out and note water levels, direction and rate of flow in rivers, streams and canals, and also flood levels and tidal and other fluctuations, where relevant.

(h) Observe and record adjacent property and the likelihood of its being affected by proposed works.

(i) Observe and record mine or quarry workings, old workings, old structures, and any other features which may be relevant.

C.3 Ground information

(a) Study and record surface features, on site and nearby, preferably in conjunction with geological maps and aerial photographs, noting as follows:

(1) Type and variability of surface conditions.

(2) Comparison of surface lands and topography with previous map records to check for presence of fill, erosion, or cuttings.

(3) Steps in surface which may indicate geological faults or shatter zones. In mining areas, steps in the ground are probably the results of mining subsidence. Other evidence of mining subsidence should be looked for, compression and tensile damage in brickwork, buildings and road; structures out of plumb; interference with drainage patterns.

TABLE 3.2
(CONTINUED)

(4) Mounds and hummocks in more or less flat country which frequently indicate former glacial conditions; for example, till and glacial gravel.
(5) Broken and terraced ground on hill slopes which may be due to landslips; small steps and inclined tree trunks can be evidence of creep.
(6) Crater-like holes in chalk or limestone country which usually indicate swallow holes filled with soft material.
(7) Low-lying flat areas in hill country which may be sites of former lakes and may indicate presence of soft silty soils and peat.

(b) Inspect and record details of ground conditions in quarries, cuttings and escarpments, on site and nearby.
(c) Assess and record, where relevant, ground water level or levels (often different from water course and lake levels), positions of wells and springs, and occurrence of artesian flow.
(d) Study and note the nature of vegetation in relation to the soil type and to the wetness of the soil (all indications require confirmation by further investigation). Unusual green patches, reeds, rushes, willow trees and poplars usually indicate wet ground conditions. (e) Study embankments, buildings and other structures in the vicinity having a settlement history.

C.4 Site inspection for ground investigation

(a) Inspect and record location and conditions of access to working sites.
(b) Observe and record obstructions, such as power cables, telephone lines, boundary fences and trenches.
(c) Locate and record areas for depot, offices, sample storage, field laboratories.
(d) Ascertain and record ownership of working sites, where appropriate.
(e) Consider liability to pay compensation for damage caused.
(f) Locate a suitable water supply where applicable and record location and estimated flow.
(g) Record particulars of lodgings and local labour, as appropriate.
(h) Record particulars of local telephone, employment, transport and other services.

TABLE 3.3
TYPICAL SPACINGS AND DEPTHS OF PITS AND BORINGS

NEW SITE OF FAIRLY WIDE EXTENT

Much will depend on the size and shape of the site, the complexity of the topography and ground conditions, and on the nature of the project. As a general guide, between 5 and 10 pits or borings should give an indication of the overall conditions on the site. Extra holes may be required to infill information where ground conditions are complex and to provide more detailed coverage where particular structures are located. When planning borehole or pit positions in relation to the locations of proposed structures it must be remembered that the proposed layout may be altered after the investigation has been completed.

FOUNDATIONS FOR STRUCTURES

For small, low-rise buidings on good ground conditions, pits or borings giving overall coverage of the site, as described above, are usually sufficient and individual buildings do not require special investigation. For larger low or medium rise buildings, groups of buildings on a site with scattered development or buildings on poor or variable ground conditions, between 2 and 4 holes will usually be needed at each building or group. Trial pits should be located outside the floor areas of the buildings themselves. Large, heavy structures with complex foundations may require more extensive investigations. At bridge sites, one hole at each abutment or pier is the minimum requirement for the simplest structures and 2, 3 or more holes may be required.

The depth of exploration should generally be at least one and a half times the width of the loaded area. Depending on the ground conditions encountered, holes may need to be deeper (in soft ground, for instance) or not so deep (such as where bedrock is encountered at shallow depth). For pile or raft foundations, closely spaced pad or strip foundations or buildings where the floor loading is significant, the loaded area should be taken as the plan area of the structure. Where the spacing of foundations is greater than 3 times their breadth, the loaded area may be taken as the width of the foundation. The depth should be measured from the bottom of the foundation or pile group.

STABILITY OF SLOPES

Between 3 and 5 borings along a critical section are usually required; more if complex ground conditions are encountered. Several sections may need to be investigated. Borings should extend well into firm ground below any possible slip surface.

ROADS, RUNWAYS AND PIPELINES

Pits or borings are typically spaced at 50m to 1km intervals, depending on the project. Major roads through complex conditions, and runways, require closer spacing of holes than lightly-trafficked country roads or pipelines. Depths should typically be 2m to 3m below formation for roads, 6m below formation for runways and 0.5m below invert for pipelines.

NOTE: the investigation requirements for each site and project combination are likely to be unique and these notes should be used only as a rough guidance.

Where good conditions are encountered at shallow depths, borings should be taken to a depth where the possible presence of weaker material, below the depth explored, would not seriously affect the proposed structure. Where bedrock is encountered, borings should extend typically about 1.5m into sound rock and 3-5m into weathered rock, though this will depend on site conditions and will be inadequate, for instance, where old mine workings may be present. At least one boring should extend well below the zones normally investigated, as a check on conditions at depth.

PITS AND BORINGS, COMMON METHODS

Trial pits

Trial pits generally provide the cheapest method of exploration and the clearest picture of soil conditions, but can only be dug to about 5m depth, at most. Pits are usually dug by mechanical excavator: the typical wheeled back-acter found on most construction sites can dig up to 10 pits a day to about 2.5m depth, depending on conditions. Larger, tracked machines are needed for greater depths, increasing costs. Pits can be hand dug down to 5m or more but the work is very slow after about 1.5m depth, because spoil cannot be thrown directly out of the pit and has to be hauled out by bucket. Nobody should enter a pit, either to dig it or to inspect it, unless the sides are clearly stable, and shoring may be needed. This slows down the operation and increases costs but it should be remembered that, if a pit collapses, anybody in it may be buried under a cubic metre or so of soil, which can weigh 2 tonnes; more than a large car. Pits cannot be dug in silts or sands below the water table or in soft clays because the sides will collapse, endangering the machine and its operator.

Disturbed, undisturbed and hand-cut samples can be taken and simple tests carried out in the sides and base of the pit. However, if the excavator is held up while testing proceeds, the number of pits dug in a day will be reduced. Variability of conditions is much more apparent in trial pits than in boreholes. The sides can be photographed for permanent record.

Rotary auger boring

This is most commonly encountered in small investigations in the form of hand augering, using the traditional post-hole auger as illustrated in Figure 3.4. Hand augers are useful in soft to firm clays

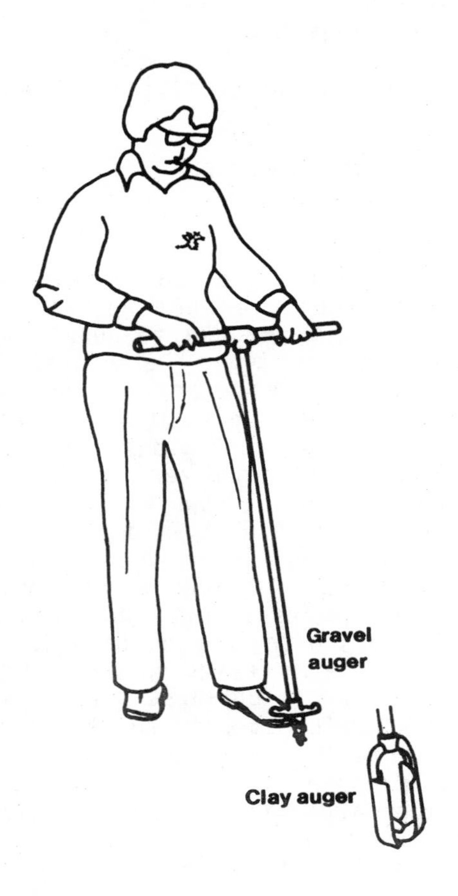

Figure 3.4 Post-hole auger

where they can reasonably be used up to about 5m deep, with hole diameters of 50-100mm, but hand augers are easily stopped by stones or even gravel. Power augers vary from the simple type illustrated in Figure 3.5 to trailer-mounted or truck-mounted augers capable of penetrating up to 50m.

Figure 3.5 Small power auger

Augers are suitable for cohesive soils, where rapid penetration rates can be achieved, but holes must be self supporting because borehole casing cannot normally be used; so they are not suitable for very soft clays or for silts or sands below the water table. Disturbed samples can be obtained from the auger but it can be difficult to identify the depths at which changes in strata occur when power augers are used. Small undisturbed samples can be obtained by hand driving tubes into the bottom of the hole. Simple in-situ testing can also be carried out.

Shell and auger boring

This uses a set of tripod legs with a power winch as illustrated in Figure 3.6 (a). Progress is achieved by repeatedly dropping an auger, consisting of a steel tube, shown in (b), to the bottom of the borehole. Clay sticks inside the auger, which is lifted out of the hole every few blows to be cleaned out, typically with a crow bar. Several other forms of auger are available for clays, such as the auger shown in (f). In sands and gravels a valve is fitted to the lower end of the tube, as shown in (c): this is a shell, or sand auger. When it is dropped into the bottom of a borehole partly filled with water, soil and water are forced through the valve which traps sediment inside. Every few blows the shell is lifted clear and emptied out. Boulders and moderate depths of rock can be penetrated by the crude but reasonably effective technique of repeatedly dropping a drilling bit or chisel down the hole ((e) and (g)). Steel lining tubes (casing) as shown in (d) are usually inserted to support the sides of the borehole. Because of the wide variety of tools available, the technique can be used in all types of soil, above and below the water table.

In clays, undisturbed samples are usually obtained using a 100mm diameter open-drive sampler. Sampling is difficult or impossible in sands and gravels so an in-situ test - the standard penetration test (described later) - is used to provide an estimate of the soil properties. Boreholes can be taken to about 30m depth in most soils but may reach 50m in easy drilling conditions.

The main disadvantages are that the method requires skilled operators and is fairly slow: a 30m borehole can take 2 or 3 days to complete. Also, it is often advantageous to add water to aid progress or help push down the lining tubes and this can affect test results or samples if excessive amounts are used.

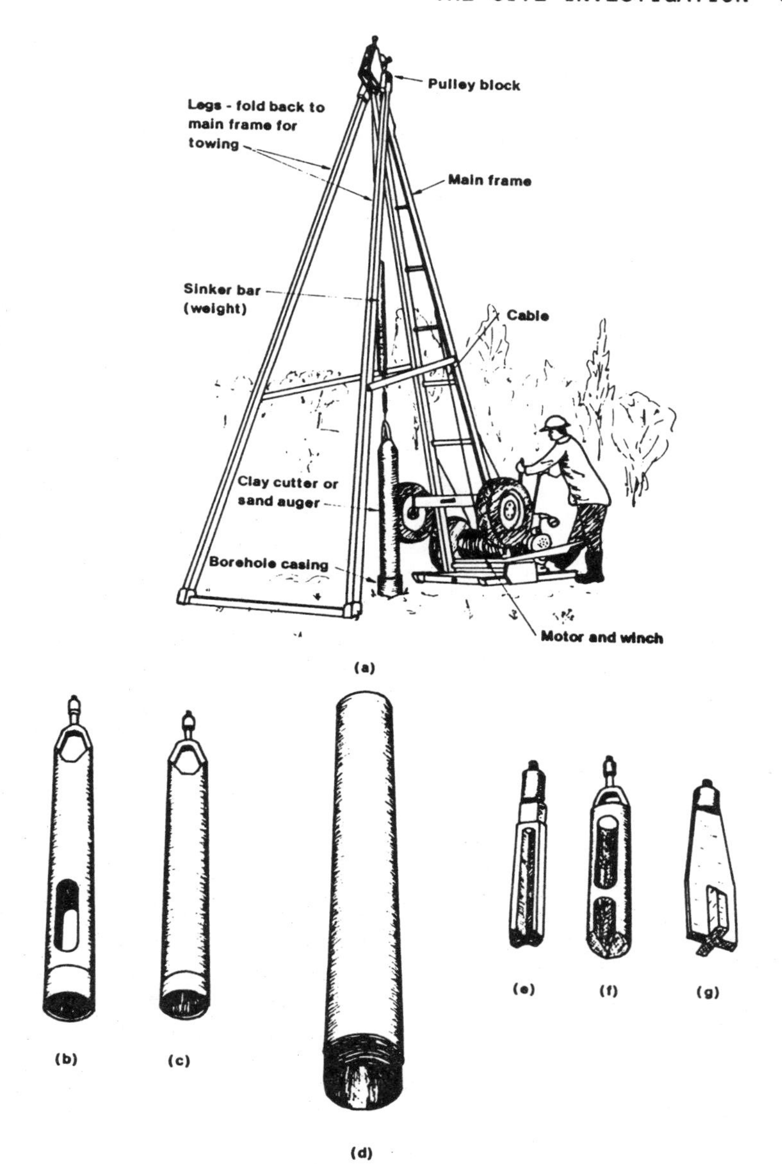

Figure 3.6 Shell and auger equipment: (a) rig; (b) clay cutter; (c) shell or sand auger; (d) borehole casing; (e) drilling bit or chisel; (f) clay auger (g) cross-head chisel.

Rotary coring

A drilling bit (Figure 3.7 (a)) with a diamond or tungsten carbide tip, is screwed on to the end of a hollow core barrel, and is rotated in the ground to produce a core of rock which is retained in the core barrel as shown in Figure 3.7 (b). A flushing medium is used to cool and lubricate the bit and to carry away cuttings. Water is the usual flushing medium but air, drilling mud or foam may also be used.

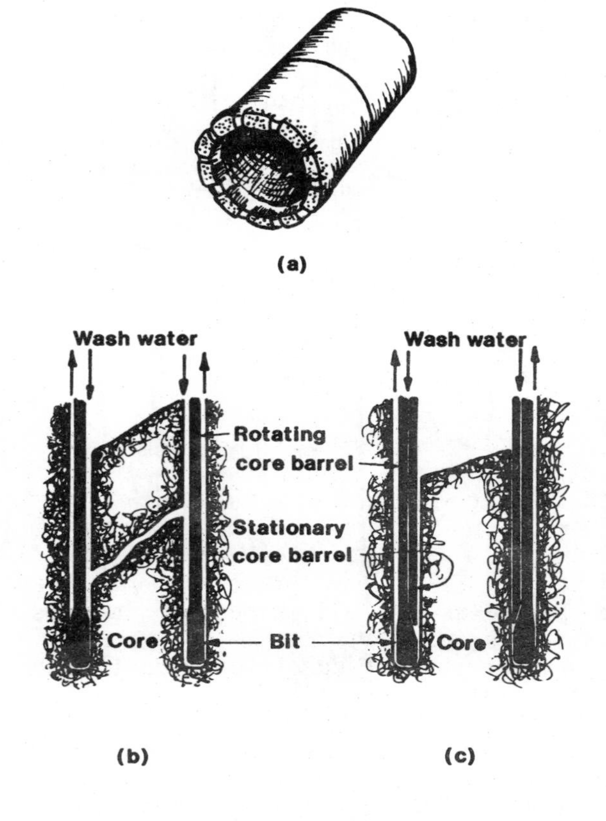

Figure 3.7 Rotary coring: (a) typical bit; (b) and (c) the principles of single- and double-barrel coring.

Rotary coring is carried out in rocks and hard clays. In the weaker materials, better quality cores can be obtained by using larger diameter bits. A further increase in quality can be obtained by the use of double-core barrels and special bits which lead the wash-water to the cutting edge without forcing it past the core sample, reducing erosion of the sample (Figure 3.7 (c)). This is important because it is the weaker, more easily eroded material which determines the overall strength of a rock mass. Casing is required to keep the hole open in weaker materials. There is a wide choice of drill sizes and types, bit types and core barrel designs. Good recovery of samples depends on the correct choice of all these, plus the speed of rotation of the bit, the pressure on it and the amount of flushing medium used. The skill of the operator is therefore an important factor in determining the quality of core samples produced.

Rotary coring can be used in conjunction with shell and auger boring or rotary drilling.

SAMPLING AND FIELD TESTING

Frequency of sampling and testing

Types and spacing of samples depend on the material encountered and the type of project. As a general guide, undisturbed samples in clays, or standard penetration tests in sands, should be carried out at 1.5m to 3m intervals and at every change in stratum, in shell and auger borings. Standard or cone penetration tests should be carried out every 1.5m in rotary drill holes through sand and gravel. Disturbed samples should be taken in all kinds of borings at 1.5m intervals and at each change of stratum. In soft clays, or for special conditions, continuous sampling may be necessary.

Excessive use of water to advance borings in clays should be avoided and, before a sample is taken, the bottom of the borehole should be carefully cleaned out.

Undisturbed samples, usually contained in sampling tubes, should be sealed with wax to prevent moisture loss. Bulk disturbed samples are normally stored in heavy-duty polythene bags tied up tightly with string.

Small disturbed samples, usually taken from the cutting shoe of an open-ended sampler or from the split-spoon sampler used in the standard penetration test, are kept in jars, tins or small polythene bags. Water samples should be taken whenever water is encountered during drilling. Samples are stored in jars whose lids are sealed by dipping them in paraffin wax.

All samples must be clearly labelled, with labels both inside and outside the containers, and must be carefully transported and stored. Samples may be discarded only when they are no longer required for inspection or testing. Care should be taken that they are not discarded too soon and all the people who may wish to make use of the samples should be informed before they are disposed of.

Open drive samplers

Undisturbed samples are usually obtained from boreholes by use of open drive samplers. The most common type is the 100mm (nominal) diameter sampler as shown in Figure 3.8 (a) and (b). Smaller samplers, usually 38mm diameter (Figure 3.8 (c)), are sometimes used for hand operations, usually in hand auger holes or trial pits. The 100mm diameter undisturbed sampler is usually referred to as a U100 but is sometimes still called a U4 after its nominal diameter in inches. In a similar manner, the 38mm diameter undisturbed sampler is referred to as a U38 or U1½. The cutting shoe protects the end of the 100mm sampler and the driving head connects it to the sliding hammer and drilling rods. The ball valve allows air and water to be expelled during driving but stops their re-entry during withdrawal of the sampler. The number of blows required to drive the sampler its full length is recorded: this gives a rough indication of the state of the ground at the base of the borehole. Care must be taken not to overdrive; that is, not to drive the sampler into the ground beyond its full length. Drillers are sometimes tempted to do this in an effort to obtain a full sample. To increase the chances of retaining a sample, the sampler should be left for a few minutes before withdrawal, to give the sample time to swell inside the tube. Also, two or three sample tubes can be connected together end to end: this is particularly useful for

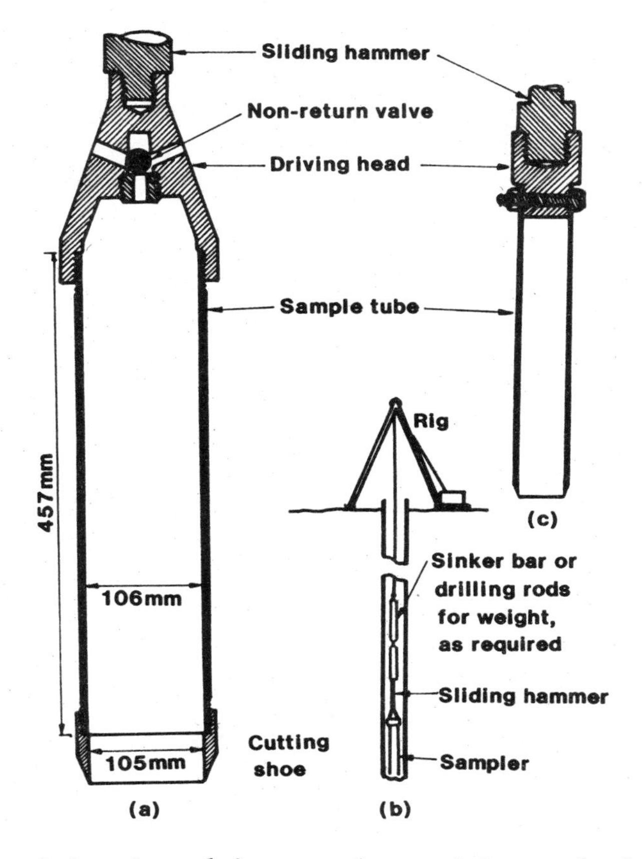

Figure 3.8 Open-drive samplers: (a) standard 100mm diameter sampler; (b) arrangement in borehole; (c) 38mm diameter sampler for hand operation.

retaining samples in soft clays and clayey sands.

The 100mm diameter sampler may be used in trial pits by driving with a sledge hammer, using a driving head and a block of wood to protect the top. However, a better method, which reduces sample disturbance, is to push in the sampler with the bucket of a mechanical excavator.

The sample is kept in the tube, protected by wax and caps at each end, until required for testing.

Piston samplers

Piston samplers are usually used in soft deposits and are pushed into the ground by hydraulic jack, rather than being driven, to reduce sample disturbance. The principal features of the piston sampler are ilustrated in Figure 3.9 (a). The sampler and piston are pushed through soft soils to an undisturbed layer with the piston positioned at the bottom of the sample tube, as shown in (b). The sample tube is then pushed into the ground, with the piston kept at a constant level, as shown in (c). Piston and tube are then withdrawn together: contact of the sample with the piston helps keep it in the tube.

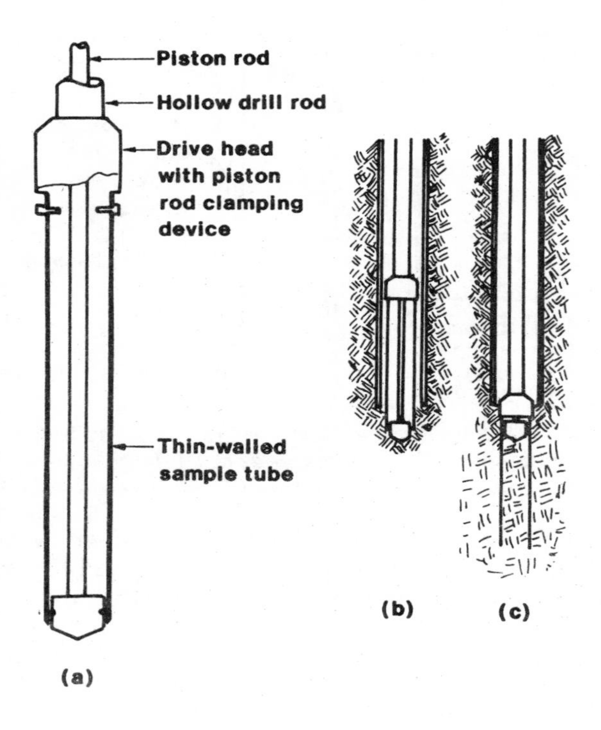

Figure 3.9 Piston sampler: (a) basic features; (b) and (c) operation in borehole.

The Standard Penetration Test

The Standard Penetration Test (SPT) is primarily used to assess in-situ properties of granular soils which cannot be sampled in an undisturbed state. It is also used in cemented soils and some clays. Before starting the test, the bottom of the borehole must be carefully cleaned out to remove any disturbed material. In sands below the water table, the sand may have a tendency to flow up into the borehole if the casing is not sufficiently far advanced, giving an unrealistically low value of density. If the casing is advanced too far, the sand below the borehole may be compacted. Thus, unless boring and testing are carried out carefully, the results of the test may be misleading.

A standard split-spoon sampler, shown in Figure 3.10 is driven 450mm into the soil by repeated blows from a hammer or "monkey" of standard dimensions. The arrangement of monkey and sampler is shown in the figure. The blows required to produce the first 150mm penetration (termed the seating blows) are usually ignored and the number of blows required to drive the sampler a further 300mm is recorded as the "N-value". It is usual, when carrying out this test, to record the number of blows for each 75mm penetration so that variations in soil conditions during driving become apparent. The standard open-ended sampler which yields a small disturbed sample is used in most soils. A solid conical end piece has to be used to prevent damage to the end of the sampler in gravel or soft rock and no sample is obtained.

Interpretation of the test is based on experience and on correlations suggested by various researchers. The various corrections and correlations commonly used when interpreting the test are given in Chapter 11. When using the test in cemented soils or weak rocks these correlations are no longer valid and any interpretation must be made with caution.

Probes

Probes measure the resistance of the ground to a rod or cone which is forced into the soil. In a sense, the Standard Penetration Test is a form of probe, but it lacks the essential feature of other probes which are

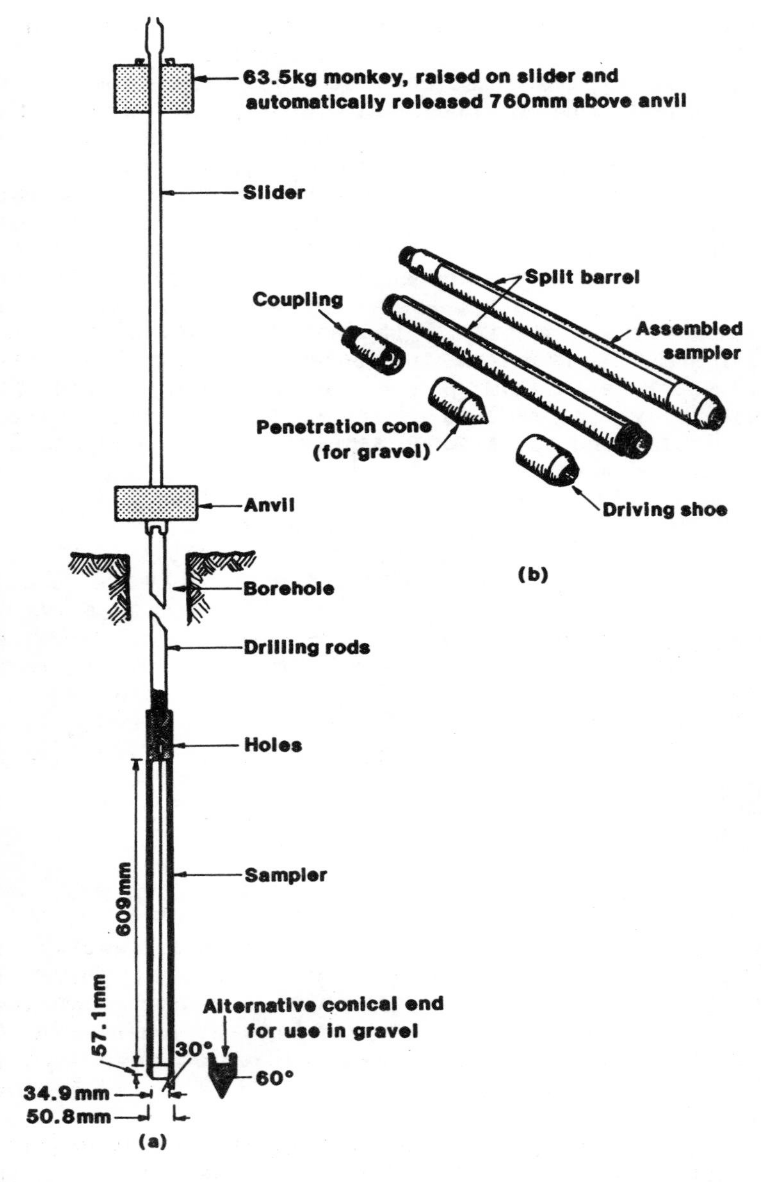

Figure 3.10 Standard penetration test equipment: (a) basic features and general arrangement; (b) details of the split-spoon sampler.

able to penetrate the ground without the need for a borehole. There are two main categories:

a. Dynamic cones, in which the probe is driven into the soil by means of a falling hammer. For penetration without the use of a borehole, it is necessary to reduce skin friction between the soil and the rod being driven into the ground. Various methods are used to overcome the problem of skin friction, as illustrated in Figure 3.11. Dynamic cone test results are usually converted to equivalent Standard Penetration Test N-values. The relationship between the cone penetration test value N_c and the SPT N-value depends on the equipment used. For the cones illustrated, the relationship $N_c = 1.5N$ has often been quoted, for depths up to about 9m.

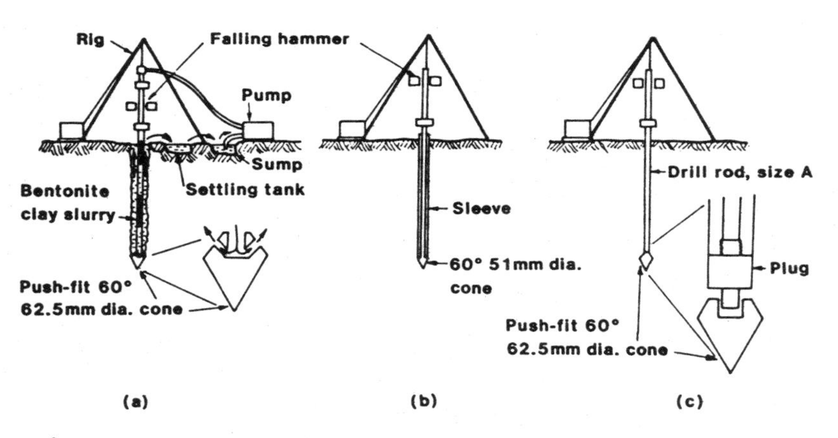

Figure 3.11 Examples of dynamic cones: (a) wash-point cone using bentonite slurry to reduce wall friction; (b) sleeved cone; (c) dry cone using a cone of greater diameter than the drilling rods to reduce wall friction. With types (a) and (c) the cone is left in the ground when the rods are withdrawn.

b. Static cones, which are jacked into the ground at a steady rate. Cone resistance and skin friction are measured separately, usually by providing a separate sleeve and incorporating strain gauges into the sleeve and tip. Results can be correlated with bearing capacity and settlement factors for foundations. An example of a static cone is shown in Figure 3.12.

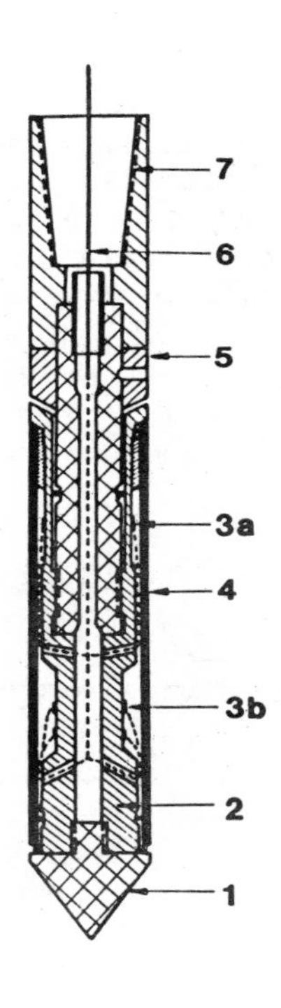

Figure 3.12 Example of a static cone. 1: conical point. 2: load cell. 3: strain gauges. 4: friction sleeve. 5: adjustable ring. 6: cable. 7: connection with rods.

A small hand probe, the Mackintosh probe, consists simply of a standard probe head and connecting rods. Resistance of the soil is measured by counting the number of blows of a standard drop hammer which are required to drive it a set distance (usually 150mm). The device is useful to give a rough indication of subsoil conditions, usually during preliminary exploration.

A fairly recent development which may be thought of

as a probe is the pressuremeter. A rubber membrane is expanded against the soil and the stress-strain characteristics of the soil are measured. Pressuremeters may be used in boreholes or may be self boring.

Vane shear testing

This test is used to measure the in-situ undrained shear strength of clay and is particularly suitable for use in soft or sensitive clays or silts which are difficult to sample or whose properties are significantly altered by normal sampling methods.

The equipment consists of a four-bladed vane, as illustrated in Figure 3.13. The length of the vane assembly is usually twice its total width: typical dimensions are 150mm long by 75mm wide for soft clays and 100mm long by 50mm wide for firm clays. The vane is inserted into the ground on the end of a rod and slowly rotated, using a special instrument to measure the torque required.

Field vane equipment is available for use either at the bottom of a borehole or for direct penetration into the ground. Smaller laboratory versions are also available.

Penetrometer tests

A penetrometer consists basically of a rod which is pushed slowly into the ground, up to a certain mark. The force is applied through a spring so that the required load can be measured by compression of the spring. By calibrating the penetrometer readings with laboratory results for a given soil, approximate values of undrained shear strength can be determined.

A small hand version, illustrated in Figure 3.14, can be carried in the pocket and is extremely useful on site to give the engineer a quick indication of the shear strengths of clays. Because of the small size of the plunger, several readings should be taken and values should be regarded only as useful guides, not as a substitute for laboratory testing.

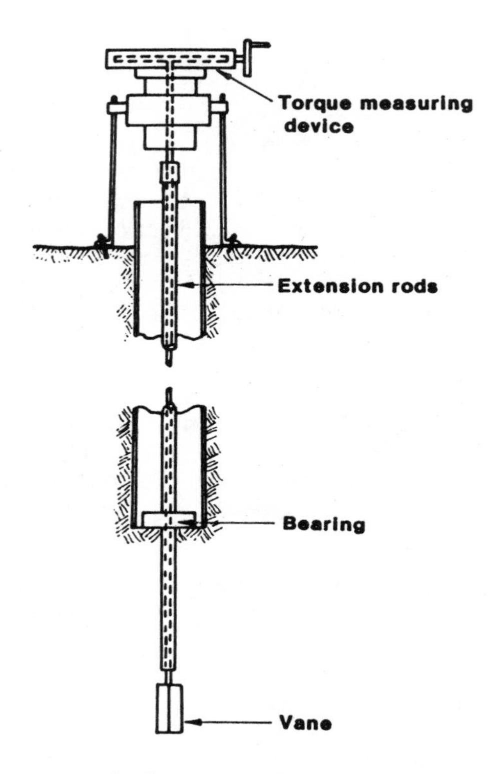

Figure 3.13 Borehole vane test.

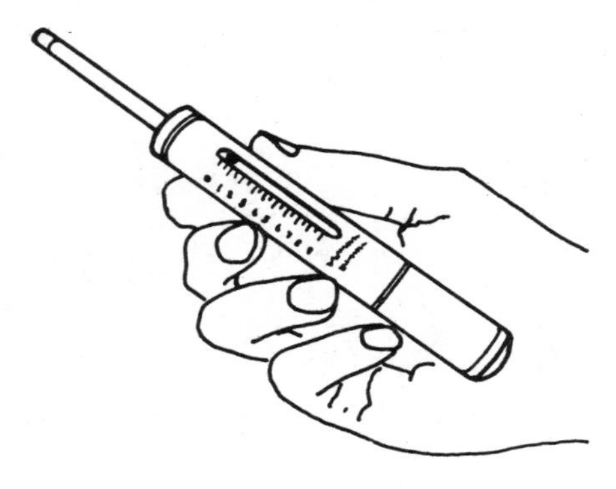

Figure 3.14 Pocket penetrometer.

Plate bearing tests

A square or circular plate is seated on the stratum to be tested, usually at the bottom of a trial pit, and loaded. The usual arrangement is illustrated in Figure 3.15. Load is applied in increments and maintained until full settlement has taken place at each increment. Plate loading tests can also be carried out in the bottom of boreholes. In some soils, special self-boring helical plates may be used allowing tests to be carried out at depth without the need for a borehole. Correlations are available to estimate the settlement of a full-sized foundation based on settlement of the plate.

Plate loading tests are particularly suitable for coarse granular materials which cannot be tested by normal laboratory means or by a penetration test. The main pitfall in predicting settlement from these tests

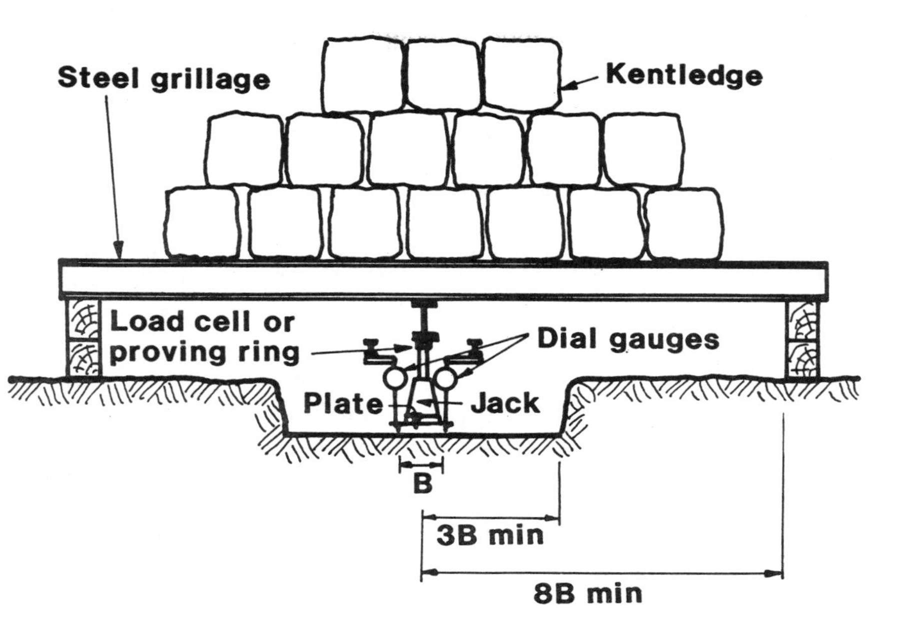

Figure 3.15 Plate bearing test, general arrangement.

is that the zone of stressed soil beneath the plate is much smaller than that beneath the larger foundation; it will thus be unaffected by deeper strata whose load-bearing and settlement characteristics may critically affect the behaviour of the foundation. With clays, tests do not usually continue for long enough for consolidation to be completed so settlement cannot be predicted. In order to obtain reliable results, plates should be as large as possible and should never be less than 0.3m wide.

Field permeability

Although laboratory permeability tests provide a straightforward method of measuring the permeability of a soil sample, they do not necessarily represent the ground permeability which depends greatly on the stratification and large scale structure of the soil. Separate field tests should be carried out to obtain true field values. A variety of test methods are used, to suit soil or rock conditions. The main disadvantages of field tests are the high costs and the problem that interpretation of the results depends on the subsoil conditions, which will be imperfectly known.

Results from field permeability tests can be used in conjunction with laboratory consolidation tests to predict rates of settlement. In some soils, especially fissured clays and stratified estuarine deposits, calculated rates of settlement based on laboratory test specimens are much too low, but those based on field permeability are more realistic.

Piezometers

It is sometimes necessary to monitor the levels and pressures of the ground water at a site and this is most usually done by installing a standpipe piezometer. The equipment consists simply of a porous pot connected to a vertical pipe which is installed in a borehole or driven into the ground as shown in Figure 3.16. Water levels are read by lowering a dipmeter as illustrated: the tip is lowered down the standpipe until a buzzer sounds or a light comes on. Graduations on the dip-meter wire enable the depth to the tip to be obtained.

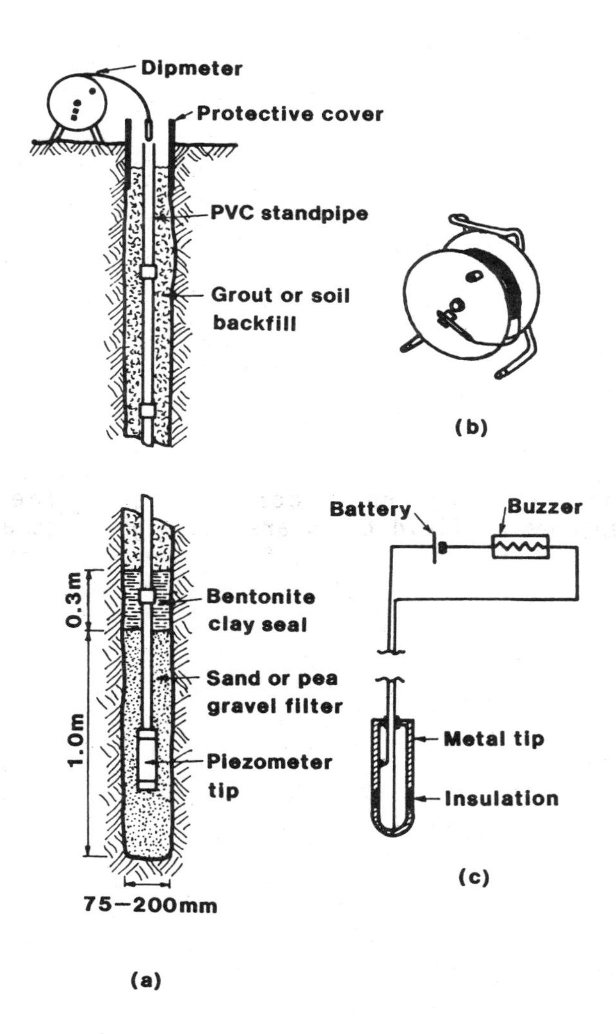

Figure 3.16 Standpipe piezometer: (a) general arrangement for piezometer installed in borehole; (b) dip meter, used to locate water level in borehole; (c) operating principle of dip meter.

Standpipe piezometers are simple, reliable and cheap but the dipmeter method of reading is slow when many piezometers need to be read. Also, the standpipes may be damaged by construction traffic or by ground movement in landslides. To overcome these problems, hydraulic piezometers are used. These are connected to pressure gauges in a central gauge-house by flexible polythene tubing containing water. More recent developments are pneumatic piezometers and electric piezometers, which use pneumatic or electric pressure transducers in the porous tip, connected to a central readout unit by gas-filled tubing or by electric wires.

GEOPHYSICAL METHODS

Geophysical surveys can be used to explore areas in greater detail and much more rapidly and economically than is possible by use of borings. Information from geophysical techniques indicates average conditions over some distance rather than the restricted vertical line of a boring, but surveys need to be carried out in conjunction with traditional borehole and trial pit investigation before the geophysical measurements can be interpreted.

The accuracy and range of soil properties obtainable from geophysical surveys is limited but the techniques are often useful for locating the boundaries of strata between borings, thus reducing the amount of costly boring required. Geophysical surveys and the interpretation of the data obtained are usually carried out by specialist firms in consultation with the geotechnical engineer.

Resistivity

A system of electrodes is used to measure the apparent resistivity of the ground, as illustrated in Figure 3.17.

A current is passed through the ground between current electrodes A and D and the potential drop between voltage electrodes B and C is measured. Usually all four electrodes are spaced evenly apart, as shown. By altering the spacing, L, the apparent resistivity of the ground will change, depending on ground conditions, and a plot can be obtained of apparent resistivity against electrode spacing. This is then matched against standard curves of idealised conditions.

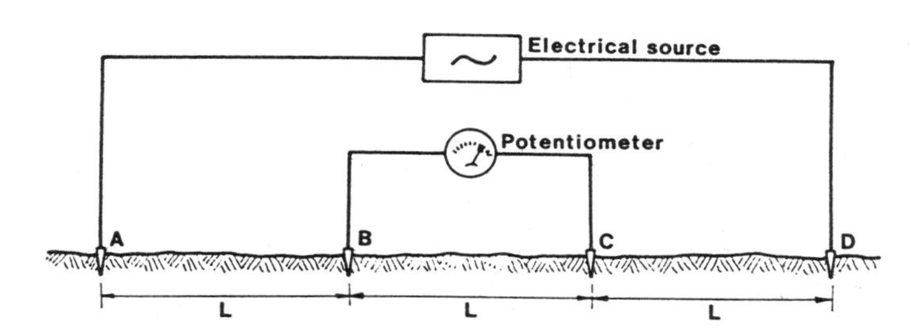

Figure 3.17 Resistivity survey; general arrangement.

An alternating current is used so that results will be unaffected by stray electrical currents in the ground. The frequency is kept low (10-100hz) to avoid inductance or capacitance effects. Voltage is measured using a potentiometer circuit.

The technique provides an inexpensive method of investigating simple ground conditions and is used to detect both vertical and horizontal variations in ground conditions. However, interpretation of results does not always give a unique solution and results may be rendered invalid by the unsuspected presence of metal pipes or cables.

Gravimetric

An instrument which is essentially a very delicate spring balance is used to measure small changes in the earth's gravity. Because of the minute variations involved, this method is usually restricted to the location of such features as large faults, ore bodies, or the edges of infilled quarries. The more accurate instruments now available are capable of detecting old mine shafts and other cavities on some sites.

Magnetic

This technique relies on the detection of small changes in the earth's magnetic field. In civil engineering, its use is usually limited to the detection of buried metal objects such as cables and pipelines.

Seismic

Shock waves are produced either by explosive charges or by a hammer striking an anvil. A line of geophones on the surface records the disturbance at various distances from the source of the shock. Shock waves travel through rocks in a manner broadly analogous to that of light rays through transparent materials, being reflected and refracted in a similar manner. This leads to two techniques of seismic surveying; the reflection method and the refraction method.

In the reflection method, geophones are located relatively close to the source compared with the depths of strata being investigated. Shock waves recorded at the geophones arrive both directly from the source and after reflection at interfaces between strata, as illustrated in Figure 3.18 (a). By noting the time lags between the arrivals of the various rays at a number of geophones, the depths to the interfaces can be calculated. This technique is suitable for locating

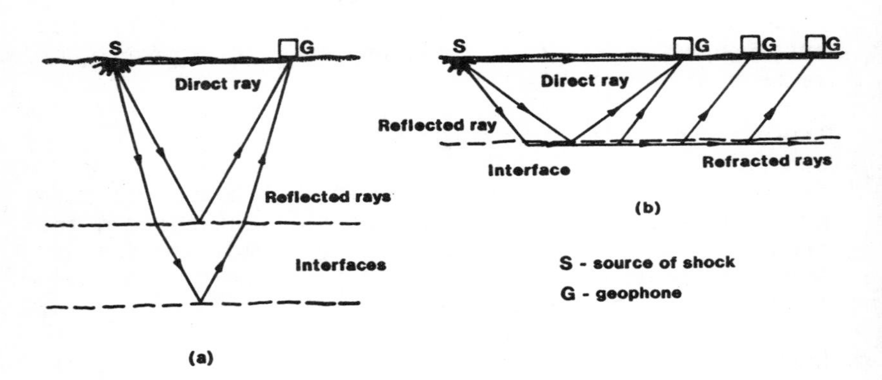

Figure 3.18 Seismic surveying: (a) reflection method; (b) refraction method.

deep formations and is more appropriate to mineral prospecting than to civil engineering.

The refraction method relies on the fact that shock waves striking an interface with a material of higher seismic velocity are deflected away from the normal (as with light rays). At a certain critical angle, the wave is deflected along the interface, as shown in Figure 3.18 (b). This critical refracted ray produces a weak head wave which is picked up by geophones, as illustrated. Close to the source, the first shock wave to arrive will be the direct wave because refracted and reflected rays must travel further. Further from the source, the first shock wave to arrive will be the refracted wave because, although it travels further, the velocity of the critical wave will be at the higher seismic velocity of the denser, underlying layer. By considering only the times taken for the first shocks to be recorded on the geophones, the depths of interfaces can be calculated. This method, which considers only "first arrivals" at the geophones, makes interpretation relatively simple and is useful for locating interfaces at fairly shallow depths. It is thus suitable for civil engineering ground investigations, particularly for locating rock head.

A limitation of the method is its inability to cope with the situation where a dense stratum overlies a less dense stratum (dense gravel overlying clay, for instance). The interface will not be recorded and the lower layers will appear to be displaced.

A recent development has been the "signal enhancement" or "signal integrating" seismograph which records and adds together the signals from a number of shock waves. The seismic shock waves, which all follow the same basic pattern, can be added together to enhance the signal, whereas any background noise, which is random, will tend to be cancelled out. By this means a seismic survey may be carried out without the need to stop plant working and a simple hammer and anvil shock source may suffice where explosives would otherwise be required.

Echo sounding and seismic reflection profiling

These techniques are used for offshore investigations. Water depth profiles can be obtained by measuring the time taken for a pulse of high-frequency sound to travel from a survey vessel to the sea bed and back. An extension of the echo sounding technique, using much higher energies, may be used to provide information on soil conditions at shallow depths below the sea bed.

Chapter 4

LABORATORY TESTS

TEST RESULTS AND THE DESIGN PROCESS

For many engineers, faced with the need to draw up a test schedule, it is not lack of knowledge of soil test procedures which causes them difficulty, but the problem of which tests to specify, and in what numbers, in order to obtain the information needed for design. Some tests are specified to obtain specific soil parameters for use in the design calculations: drained shear strength tests for slope stability or consolidation tests for settlement calculations, for instance. Other tests are included only for soil classification and it is these tests which may give the engineer most uncertainty when deciding on a testing schedule. This is a situation in which a little experience is worth a great deal of book-learning. Table 4.1 indicates how test results are commonly used in the design process and which results are applicable to various types of structures and projects.

Details of test procedures and laboratory equipment required are given in various standards and texts. The specified procedures must be carefully followed because slight variations in procedure can have a marked effect on test results. When test results are reported the standard or text used should be given along with details of any departure from the specified procedure.

Test procedures are described here only in outline, to make the reader aware of the overall technique used. Emphasis has been placed on giving an understanding of the property that each test measures and providing an awareness of errors and misleading results which arise from problems that can occur during testing.

TABLE 4.1
APPLICABILITY OF LABORATORY TEST RESULTS TO DESIGN

CLASSIFICATION OR INDEX PROPERTY TESTS

These include natural density and moisture content, specific gravity of soil particles, grading and plasticity tests. The information from these tests can be used in two ways:

a. **Directly.** For instance, density may be used in the calculation of pressure on retaining walls; grading may be used in the calculation of filter design; and grading and plasticity may be used with certain design procedures to calculate pavement thicknesses.
b. **Indirectly.** Tests which measure the strength and deformation characteristics of soil are usually expensive and require undisturbed samples, so the number of tests which can be carried out is limited. Also, the engineer must decide whether variations in test results represent random variations in the same soil type or indicate a change of strata. By supplementing these tests with larger numbers of relatively cheap classification tests on disturbed samples, along with visual examination, soils can be classified into types for the purpose of interpreting strength and deformation test results and deciding on design values.

Grading and plasticity tests are included in most testing programs: they provide useful information for direct or indirect use and are relatively cheap to carry out. Grading of the silt and clay fractions may be obtained by sedimentation analysis but this information is rarely of use: for the majority of work, sieving alone is sufficient.

Natural density and moisture content tests are carried out wherever the ground is to be utilised in its natural state (beneath foundations or cuttings etc.) and sometimes where the soil is to be treated, for comparison. These tests are routinely carried out in conjunction with triaxial, shear box and consolidation tests.

The specific gravity of the soil particles is needed to determine the degree of saturation of the soil and to enable 'e-log p' plots to be made from consolidation test results, where this is required.

PERMEABILITY TESTS

Laboratory permeability tests are appropriate for compacted soils in dams, filter materials and drainage materials. The permeability of naturally-occurring deposits is better determined from field permeability tests but laboratory tests may sometimes form a useful supplement to a field testing program (for checking the ratio of vertical and horizontal permeability, or variation of permeability with density, etc.).

The permeability of clays is usually calculated from consolidation (oedometer) tests.

COMPACTION TESTS

Compaction tests are carried out in situations where compacted soils are to be used, such as roads, airports, filled areas, land reclamation schemes and embankment dams. The most common test are the BS or AASHTO standard or heavy compaction tests. The choice of standard or heavy compaction test depends on the compaction requirement for the project and the type of plant likely to be used. These, in turn, are governed by the severity of loading and how much settlement can be tolerated.

TABLE 4.1
(CONTINUED)

CALIFORNIA BEARING RATIO (CBR)

This test was devised specifically for pavement design and is used only in conjunction with one of several empirical design procedures which are based on the use of this test. Results of CBR tests show a wide scatter and are affected greatly by sample preparation. Details of sample preparation must be precisely specified so that the test results are relevant to the design and environmental conditions of the pavement under consideration.

SHEAR STRENGTH

Shear strength parameters are required for the design of retaining walls and all types of foundations and for calculating the stability of embankments, cuttings and natural slopes. Tests fall into two main categories.

a. Quick undrained tests which give total stress parameters.
b. Consolidated slow tests either with drainage and complete pore pressure dissipation or without drainage but with pore pressure measurement.

Triaxial tests are the most common shear strength tests, using either 3 separate 38mm diameter specimens or a single 100mm diameter specimen, tested at 3 confining pressures. Where larger samples are required or testing is to be carried out to very large strains, or for granular soils, shear boxes may be used.

Practically all foundation design is based on total stress parameters using quick undrained tests which are simple and cheap. This gives the end-of-construction condition which is usually the worst case. With slope stability problems, pore pressures must be taken into account and the more expensive slow tests, which give effective stress parameters must be used.

In sensitive clays, in-situ vane testing is a better method of obtaining shear strength, to avoid sample disturbance.

CONSOLIDATION TESTS

Consolidation testing is usually carried out wherever settlement characteristics are required for fine-grained soils. Tests are usually carried out on natural soils beneath proposed foundations or embankments and on compacted samples of soils to be used for embankment construction. Tests are rarely carried out on sands because of the difficulty of obtaining undisturbed samples.

In estuarine deposits and fissured clays, the coefficient of compressibility may be reliably obtained but, because the rate of consolidation is affected by the macro-structure of the soil, values of the coefficient of consolidation are unrelaible and should be obtained indirectly from field permeability tests. Consolidation tests also give unreliable results with tropical residual soils.

CHEMICAL TESTS

These tests are used to check the acidity of the soil and the quantities of aggressive materials in the ground, such as sulphates, chlorides and organic materials which may attack buried concrete or metal.

MOISTURE CONTENT

The moisture content of a soil has a profound influence on its properties: it affects the shear strength, the consolidation properties and its response to compaction. It may also affect future swelling or shrinkage of the soil. Moisture content tests are therefore frequently carried out and are routinely included as part of the standard procedure for most tests. They can be used to determine the natural moisture content of a soil or its moisture content at some particular state such as the liquid or plastic limits, described later.

The standard procedure for the moisture content test is to heat up a known weight of soil in an oven and to find the loss in weight on drying. The sample must be heated for long enough to drive off all the water, but typically it is simply left in the oven for about 18 hours (overnight) which is usually ample time. With clay soils the moisture content will depend to some extent on the temperature of the oven. This is because some water molecules are loosely attached to the clay molecules by weak electrostatic forces to form an "adsorption complex" - a thin film of water surrounding each clay particle which can be thought of as part of the soil structure. To avoid driving off this water the oven temperature is kept to 105-110°C - hence the long drying time required. A sample weight of 10g to 30g is sufficient for a test on clay soils but for gravels 300g or more may be needed to obtain a representative value.

The moisture content is calculated as the loss in weight of the sample expressed as a percentage of the dry weight of soil.

The length of time needed to dry the sample can be a problem, particularly with road or embankment construction where frequent and rapid checking is required. To overcome this, a number of alternative methods may be used but results must be checked against the standard method for each soil type tested.

In the "sand bath" method the tin containing the sample is heated by placing it on a tray of sand which is heated by a gas burner. This avoids direct contact with the flame but heats the soil to a high temperature so that it dries out fairly quickly. The method is suitable for some soils but results are often erratic. An alternative method of heating the soil, which usually produces fairly accurate results, is to mix it with methylated spirit and set it alight. This is a process which must be attempted with care!

A popular quick method is the "Speedy" tester. A fixed weight of soil is shaken with calcium carbide powder in a sealed pressure flask. Reaction of the powder with the soil moisture produces acetylene gas. The resulting pressure generated is registered on a pressure gauge which is calibrated directly in terms of moisture content. Experience with this method shows that results must be calibrated against oven dried samples for each material type and, even then, it is suitable only for testing clean aggregates. Because it is a proprietary device, too much faith is often placed in the results without any checking.

A relatively new method, which gives excellent results with most types of soil, is the microwave oven method. An ordinary domestic microwave oven can be used and small samples take typically 10 to 30 minutes to dry out, depending on power setting and oven load. Metal containers cannot be used - glazed porcelain evaporating dishes are probably best - and it is advisable to include a dummy sample in the form of a piece of asbestos cement so that there is always something to absorb the microwaves, even when the soil has dried out.

PLASTICITY TESTS

When dry, or at very low moisture contents, clay forms solid lumps. At the other extreme, when it is very wet, it behaves like a viscous liquid. For a certain range of moisture content between these two extremes it is a plastic solid which can be moulded into different shapes. The moisture content which represents the borderline between the liquid and plastic states is known as the liquid limit. The lowest moisture content at which the clay can be moulded without cracking is known as the plastic limit. The difference between these two, the range of moisture content over which the soil is plastic, is known as the plasticity index. Plasticity is exhibited only by clays and silts, not by sands and gravels. There are, in reality, no sharply defined values at which these transitions take place so, in order to obtain repeatable results which everybody can agree on, arbitrarily defined testing standards must be used.

For the plastic limit, clay is simply hand-rolled into threads on a glass plate. The process is repeated until the threads just begin to crack: the plastic limit has then been reached and the soil is tested for its moisture content. To obtain repeatability, the diameter of the threads and the amount of rolling is specified.

To obtain the liquid limit, a paste of soil is pressed into a specially-shaped cup and a groove of standard dimensions is made through the middle of it. The cup is tapped gently and the number of blows needed for 13mm of the groove to close is recorded. The process is repeated for a range of moisture contents. The moisture content at which the soil would close up after just 25 taps is defined as the liquid limit. To ensure repeatability, the test is carried out using a device of standard dimensions which automatically raises the cup a fixed distance and allows it to drop on to a block of rubber of defined hardness and resilience, as shown in Figure 4.1 (a). The device must be placed on a hard, solid surface. If used on a flexible worktop, for instance, liquid limit values obtained will be too high. An alternative method, introduced in 1975, is to drop a standard cone into the soil paste and record the penetration, using the device shown in Figure 4.1 (b). This is repeated for a range of moisture contents and the moisture content at which the cone penetrates 20mm is defined as the liquid limit. The value obtained should be independent of the test used, but the cone penetrometer is slightly quicker, produces less scatter of results, and is unaffected by the stiffness of the working surface used. About 250g of soil is needed, after sieving out the coarse material, for the combined liquid and plastic limit tests.

Although simple, almost comical, in their definition, the tests require some skill to perform them satisfactorily, since judgement is required. Results with tropical residual soils depend on the method of sample preparation and should be viewed with some scepticism.

Plasticity tests are useful for classifying soils and many engineering properties can be inferred from the results. They are therefore carried out on a high proportion of samples obtained from ground investigations.

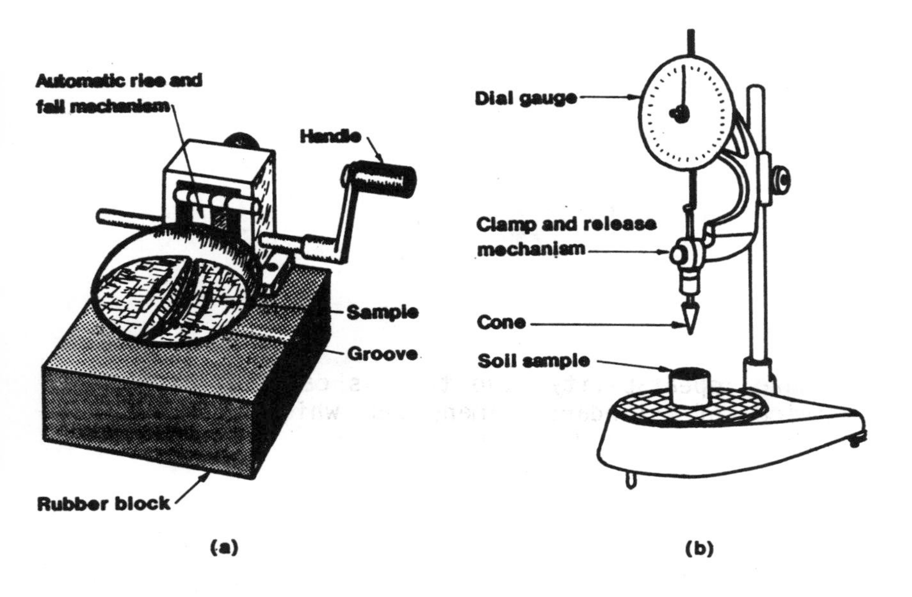

Figure 4.1 Liquid limit test equipment: (a) Casagrande apparatus; (b) cone penetrometer

SPECIFIC GRAVITY OF SOIL SOLIDS

The method used is to weigh the oven-dried soil and then shake it up with water in a container of constant volume so that the weight of water displaced can be measured. Between 20g and 400g of soil is needed, depending on the coarseness of the soil and the size of container used. Typical containers are shown in Figure 4.2.

An accurate value of specific gravity is required for most earthworks and compaction contracts to allow precise determination of the percentage volume of air present in the soil. In the majority of other cases, an assumed specific gravity value in the range 2.65 to 2.70 is sufficiently accurate.

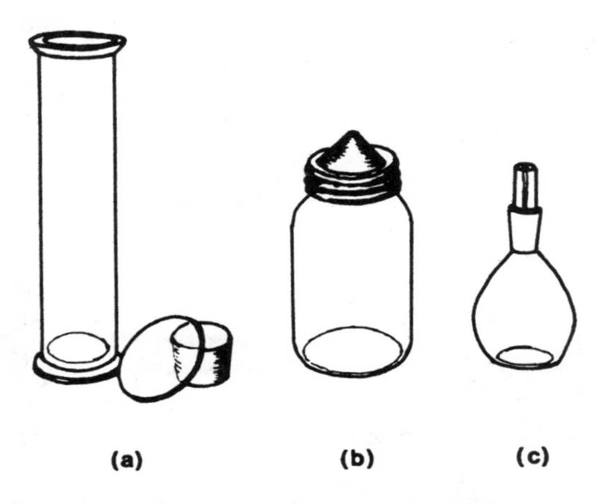

Figure 4.2 Specific gravity determination equipment: (a) gas jar with glass plate and bund; (b) pycnometer; (c) specific gravity bottle

PARTICLE SIZE DISTRIBUTION

The particle size distribution, usually referred to as the grading, is an important property of coarse-grained soils - sands and gravels. It is used for classification and to infer a number of engineering properties of these soils in much the same way that the plasticity properties are used with clays. Two basic techniques are used: sieving to separate out the various particle sizes down to fine sand size; and sedimentation analysis, which utilises the rate at which particles settle out of suspension when stirred with water, for silts and clays. Sample size depends on the coarseness of the soil: 200g is sufficient for clays but 2kg or more may be needed for gravels; up to 20kg if cobbles or boulders are present.

The sieves used are specially made so that they slot on top of each other to form a column or "nest", with mesh size becoming progressively finer the further a sieve is down the nest. Soil is placed in the topmost sieve and the entire nest is shaken, usually on a mechanical shaker. Weights of soil retained on each size sieve are recorded. This simple technique, known as "dry sieving", is satisfactory for clean sands and gravels but does not work with soils containing silt or clay, which stick together in lumps. For such soils, a method known as "wet sieving" is used. Clays and silt are first removed from the sample by washing the soil through one or two of the finer sieves. The soil retained on these sieves, which now contains only the granular material, is dried and sieved in the usual way. Wet sieving should always be specified for soils unless a particular sample is obviously granular. Dry sieving is normally specified for aggregates.

Silts and clays are too fine to be sieved so sedimentation methods are used. The soil is shaken with water in a measuring cylinder and allowed to settle. The rate of settlement depends on the particle size, as expressed by Stokes' Law. The concentration of particles at a particular depth will decrease with time but the rate of change will depend on the distribution of particle sizes in the soil. The concentration of particles at a standard depth is measured at various times. This is done either directly, by removing a small sample of the suspension using a pipette; or indirectly by lowering a hydrometer into the suspension to measure its density. The method is actually more complicated than the brief description given here because of the need to ensure complete deflocculation of the soil. Analysis depends on Stokes' Law, which assumes that the soil particles are spheres, whereas in reality clay particles are plate-like. Unless it is carried out for a very long time the test will give only the grading of the silt range and not of the clay particles. But since the plasticity properties of the silt and clay content are much more important than their grading, there is usually little point in carrying out this test. It is more often specified out of ignorance than for any valid reason and should not be included in a testing schedule unless the grading of the fine material is needed for some special reason.

The results of sieving and sedimentation tests are plotted out as a particle size distribution curve which provides a visual representation of the soil grading as indicated in Figure 4.3.

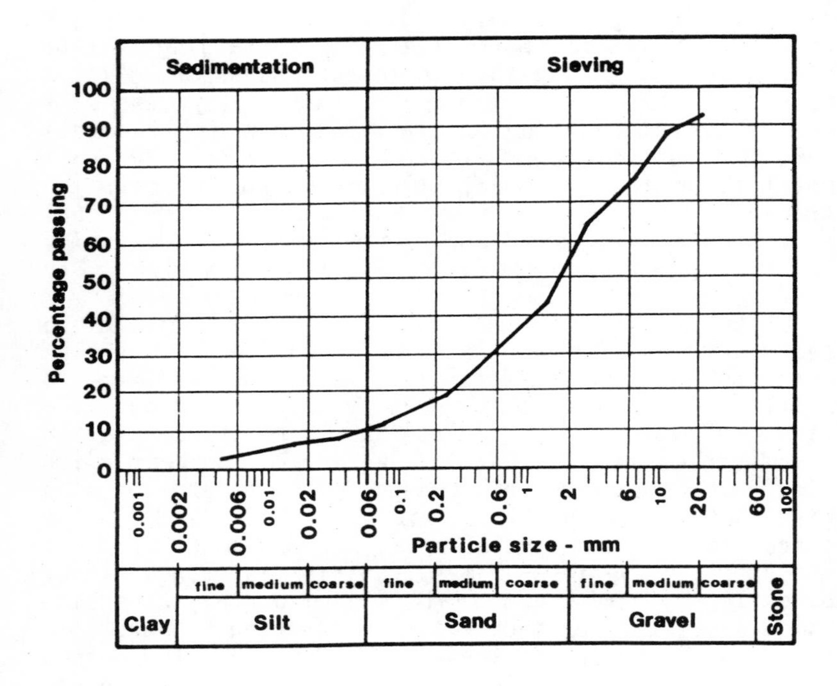

Figure 4.3 Example of a grading curve

TRIAXIAL TESTS

Triaxial tests are the most usual form of strength test for soils and they come in a confusing variety. Undisturbed specimens of soil in the form of cylinders are tested in axial compression but the specimen is enclosed in a pressure cell so that it can be surrounded by water under pressure to simulate the effects of pressure from the surrounding earth. A thin rubber sheath surrounds the specimen to separate it from the water. A typical arrangement for the test is shown in Figure 4.4. Triaxial tests are specified wherever problems involving soil strength arise. These include all types of foundation problems; tunnels;

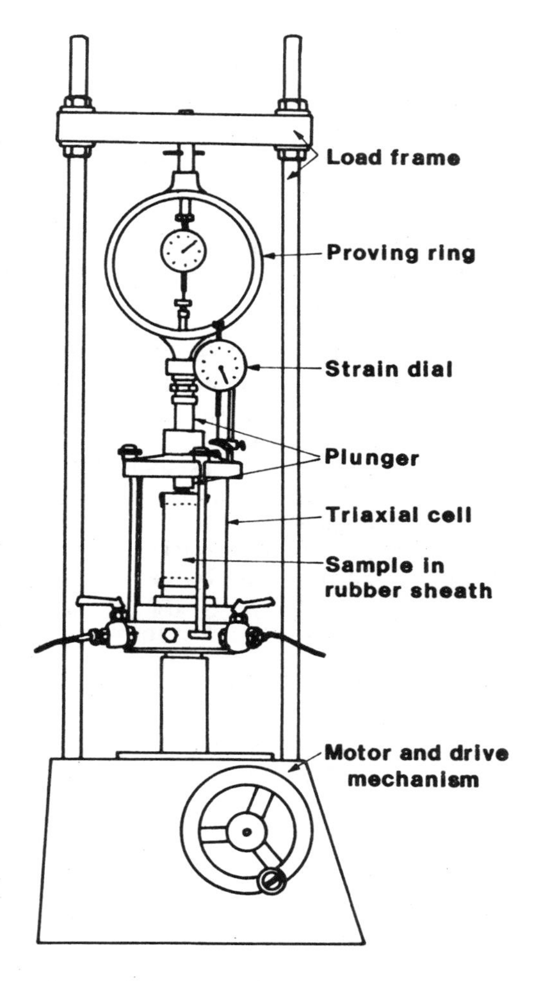

Figure 4.4 Triaxial test equipment

retaining walls; and the stability of cuttings, embankments and natural slopes.

The quick undrained test is the most unusual type of triaxial test carried out. The specimen is sealed by the rubber sheath and plastic end platens so that no drainage can take place during the test. It is loaded at a constant rate of strain, typically 2% per minute, so that the specimen reaches peak strength after about 10 minutes. Three 38mm diameter specimens, 76mm long, can be conveniently made, side-by-side, by extruding soil from a standard U100 (100mm diameter) sample tube into three 38mm diameter specimen tubes. Each of the three specimens is tested at a different confining pressure to simulate the pressure range likely to occur in the field. Choice of confining pressure depends on the particular problem being studied but for many foundation problems values used are; overburden pressure, overburden plus 100kN/m^2, overburden plus 200kN/m^2. This test is quick, easy and fairly cheap to carry out, and the test conditions simulate the immediate response of the soil to loading - the end of construction condition. Conveniently, this is the most critical condition for foundation problems.

With soils containing gravel the 38mm diameter specimens are too small and it is usual to test a 100mm diameter specimen, 200mm long, obtained by extruding directly from the U100 sample tube. The problem arises that three specimens would require more than 600mm length of sample, which is longer than a standard sample tube. Even if longer samples were obtained, there could be significant variation of the soil over 600mm, making any comparison of the strengths obtained suspect. To overcome this problem, a single specimen is tested but, as soon as it begins to fail, shearing is stopped and the cell pressure is increased to the next value. The process is repeated so that three failure strengths, corresponding to three different cell pressures, are obtained. This is a difficult and not altogether satisfactory procedure which requires a skilled technician and does not always yield good results.

The drained test is used to simulate the conditions encountered with slopes and retaining walls which usually reach their most critical condition some time after construction, when excess pore water pressures in the soil have had time to dissipate and consolidation has taken place. Porous discs and special platens on the ends of the specimen allow drainage and consolidation. The test is carried out in two stages. First the specimen is left in the cell and allowed to consolidate under the influence of the cell pressure. Once this is complete, the specimen is sheared. The rate of shearing is kept low to eliminate pore water pressures which might build up during the test. Three specimens are usually tested, at different cell pressures, as with the undrained test.

The consolidated undrained test is used for certain special applications, such as the stability analysis of a dam under rapid drawdown conditions. The specimen is consolidated in the manner used for the drained test but drain taps are closed during the shearing stage. For meaningful results to be obtained from this test the rate of shearing must again be low, to allow equalisation of pore water pressures throughout the sample during shearing, and pore pressures must be measured as shearing takes place; the upheaval of shearing which takes place in the sample can generate positive or negative pressures. A variation of this test is sometimes performed in which the shearing stage is performed quickly and without any pore pressure measurement. This has the advantage of speed, simplicity and cheapness but the results cannot sensibly be used in analysis and may even lead to unsafe design.

SHEAR BOX TESTING

This is a less-used alternative to the triaxial test for shear strength testing. A block of soil, typically 60mm square by 20mm thick, is placed in a metal box, as shown in Figure 4.5. The box is split horizontally, as shown, so that the top and bottom halves can be slid across each other, shearing the sample through its middle. The box and sample are placed inside a sliding container which transmits the shearing force to the lower half of the box and allows the sample to be flooded during test, if required. The bottom of the box is pushed along slowly at a constant rate and the shearing force is measured by a proving ring or transducer. A typical rate of shear is 1.25mm per minute, continued for about 9mm. In place of the cell pressure used in the triaxial test, a confining pressure is applied through a loaded plate resting on the sample. Either clays or sands can be tested and larger boxes are available for testing gravels and gravelly soils. Drainage conditions are more difficult to control than in the triaxial test. Also, stress conditions within the soil are not as simple as they first appear, although this objection is probably more theoretical than practical for most purposes. The test can be run to much larger strains than with the triaxial test, which makes it useful for obtaining residual strengths, for landslide studies for instance. Sample preparation is inconvenient for clays, but the greatest advantage of the test is that sand and gravel specimens can be made up with ease because the specimen is contained in a box. Because of this, its most common use is for shear strength testing of sands.

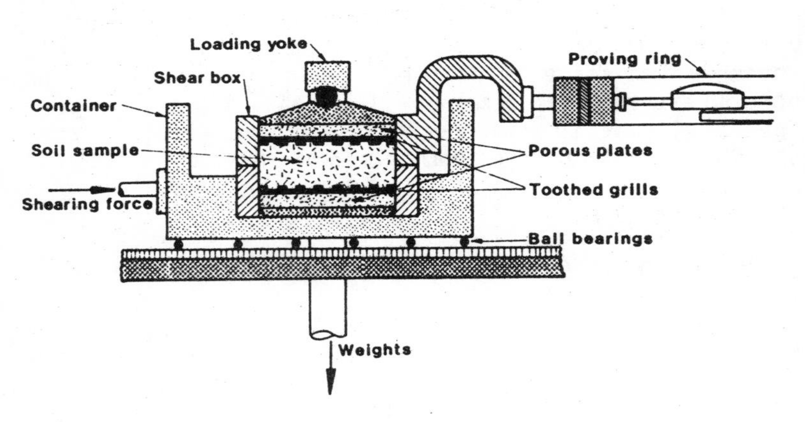

Figure 4.5 Principal features of the shear box apparatus

CONSOLIDATION TESTING

The consolidation properties of clays are usually measured using the one-dimensional consolidation apparatus, usually known as an oedometer or (in the U.S.A) a consolidometer, as illustrated in Figure 4.6 (a). A disc of (usually undisturbed) soil is cut from a sample using a special cutting ring. The usual size of specimen is 20mm by 75mm diameter, although diameters from 50mm to 100mm are used. Once it has

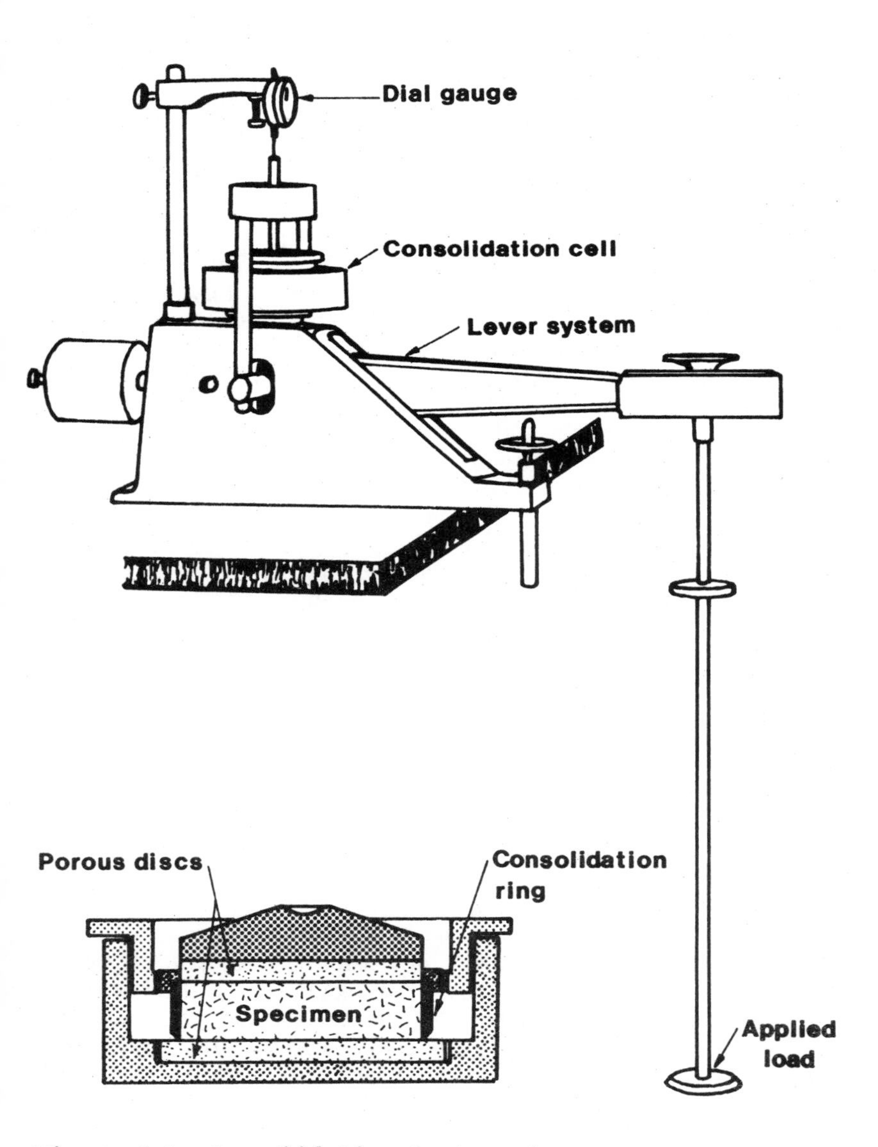

Figure 4.6 Consolidation test equipment

been cut and trimmed at each end, the specimen, still in the cutting ring, is placed in a consolidation cell. This is shown diagramatically in Figure 4.6 (b). The cell is placed in the consolidation machine and a load is applied via a loading yoke and lever system. A stop-clock is started the instant the load is applied and it is usual to flood the specimen by pouring water into the consolidation ring as soon as possible after the load has been applied. A dial gauge records changes of thickness of the specimen. Dial gauge readings are typically taken at 0, ¼, ½, 1, 2, 4, 8, 15 and 30 minutes and then at 1, 2, 4, 8 and 24 hours. It is usual to start the test in the morning, leave the sample under load overnight and take a final reading the next morning, so that a complete load cycle takes 24 hours. Further weights are then added to the loading system and the process repeated. Usually the test is carried out for 4 load increments. Often, after consolidation under the final load is complete, all weights are removed and the specimen is allowed to swell for at least 24 hours. The weights used are usually calculated so that loading pressures are selected from the sequence 25, 50, 100, 200, 400, 800, 1600, 3200kN/m^2 (0.25, 0.5, 1, 2, 4, 8, 16, 32 tons/ft^2), the choice of loads depending mainly on the expected stresses in the soil, including overburden pressures. With 4 load increments and the swelling stage, the test usually takes a working week.

Two parameters are obtained from the test: the coefficient of volume compressibility, m_v, which can be used to calculate the total amount of settlement; and the coefficient of consolidation, c_v, which can be used to calculate the rate at which settlement is likely to occur. High orders of accuracy should not be expected from settlement calculations. To underline this point, predicted settlements are usually given using such terms as "less than 25mm" or "between 100mm and 150mm". Rates of settlement are subject to even more vagueness, depending on the soil type. For most soils, reasonably realistic estimates of settlement time can be made using c_v values but for some types of soil the predicted times will greatly exceed the actual settlement times. This is especially true of estuarine clays containing silt and sand layers and of fissured clays. A factor of 10:1 or more between predicted and actual consolidation times is not uncommon for these soils so that where settlement is predicted to take 3 years, for instance, it may actually be substantially complete after only 3 months. This is because the rate of settlement depends on the overall permeability of the soil mass but, when soil is tested in a consolidation apparatus, the large-scale features such as fissures or sand lenses are not represented in the small specimen. The measured c_v value does not therefore represent the true value for the site conditions. This error can be important: were settlement to be substantially complete within a few months, then finishes to a building or surfacings to a road embankment could be delayed until most of the movement had taken place. However, this advantage cannot be taken unless the rate of settlement can be accurately predicted. Where necessary, settlement times can be more accurately predicted by carrying out field permeability tests, in boreholes, to measure the overall permeability of the soil mass.

SOIL COMPACTION TESTS

These are used wherever compacted earthworks are required; on road construction and land reclamation projects, for instance. The test produces two values; a soil density and a moisture content. The density value represents a reasonably achievable density which will give a well-compacted soil. It is used as a standard against which field densities may be judged, to assess whether earthworks have been adequately compacted. The moisture content value represents (at least as a good approximation) the best moisture content for the soil to have during compaction to obtain the maximum benefit from the compactive effort used. The soil density achieved can be expressed in two ways: bulk density which includes the weight or mass of both solids and water; and dry density which considers only the soil solids. Dry density is the more appropriate measure of compaction since it represents the state of packing of the soil solids.

A sample of air-dried soil is passed through a 20mm sieve and mixed with a little water; the amounts of water required during the test must be judged from experience. A layer of soil is placed into a standard mould, with baseplate and collar attached, as shown diagramatically in Figure 4.7. It is compacted by

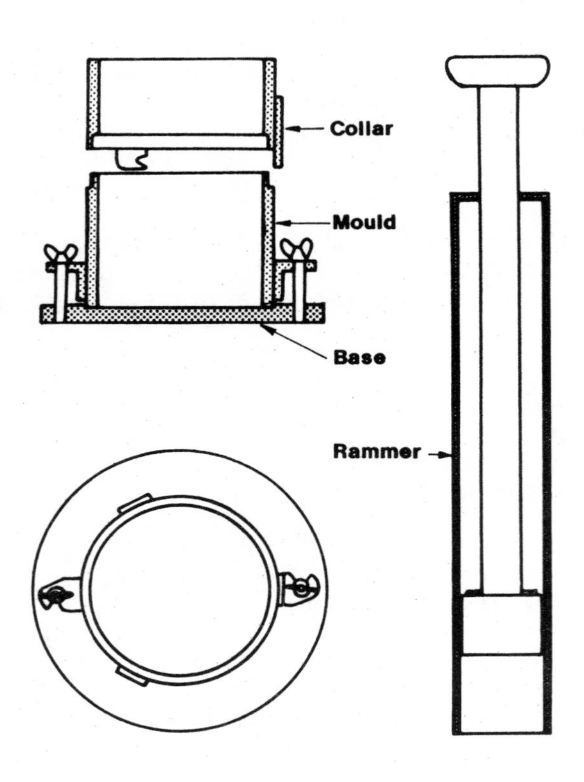

Figure 4.7 Typical compaction mould and hand rammer used in compaction tests

repeatedly dropping a rammer onto the surface. Further layers of soil are added until the mould is slightly over-full; compacted soil protruding slightly into the collar. A standard amount of compactive energy per unit volume is achieved by specifying the size of the mould, the number of layers, the weight and fall of the rammer and the number of blows per layer. The collar is then removed and the excess soil is struck off level with the top of the mould using a straight edge. The mould is now weighed to obtain the density of the soil and small specimens are taken for moisture content testing. The test procedure is repeated for various moisture contents: typically 5 or 6 values are obtained. From the results, a graph can be drawn to obtain the maximum dry density (MDD) and the optimum moisture content (OMC) at which this is achieved (see Figure 4.8). In addition to MDD and OMC it is useful

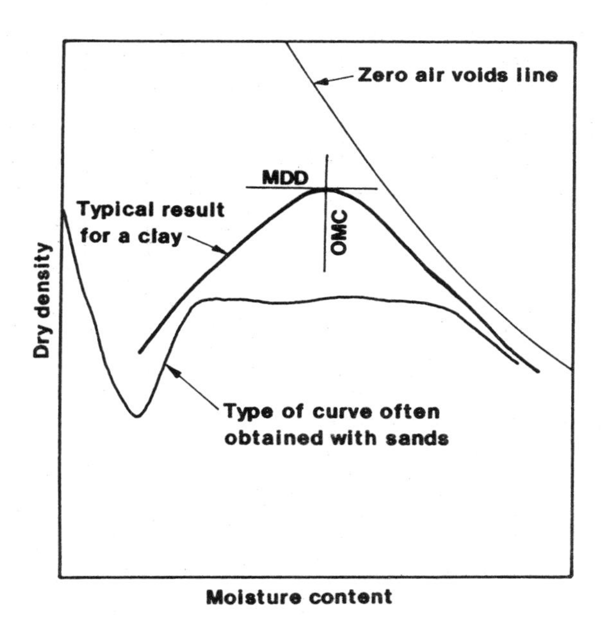

Figure 4.8 Typical dry density/moisture content relationships obtained for compaction tests

to know the amount of air present in the compacted soil, since a low air content implies good compaction. The air voids present (the volume of air expressed as a percentage of the total volume of the soil) can be calculated for any value of dry density and moisture content provided the value of the average specific gravity of the soil particles is accurately known. The zero air voids line shown in Figure 4.8 gives the dry density that would be achieved at any given moisture content if it were possible to drive out all the air in the compaction process: it is the maximum value of density theoretically achievable. This is impossible to attain in practice but, beyond optimum moisture content, the compaction curve tends to approach this limiting line.

The two most usual forms of the test in Britain both use a one-litre mould. Compaction is either in three layers using a 2.5kg rammer falling through 300mm or, to give a heavier standard, in five layers using a 4.5kg rammer falling through 450mm. In both cases, 27 rammer blows per layer is specified. A larger California Bearing Ratio mould is used for coarser soils, with a corresponding increase in the number of blows per layer. The compaction rammer used may be manual, as illustrated, but is usually electrically driven and automatically controlled. Usually the same soil is used repeatedly throughout the test, so that 5kg is sufficient. Soils containing friable material tend to break down during test so a fresh sample must be used for each moisture content. This increases the amount required to about 30kg; more if a larger mould is used. Sands and gravels do not compact well under the standard rammer and give unrealistically low values. To overcome this, a vibrating hammer is used to compact the material. Values obtained from the test are not fundamental properties of the soil but depend on the compactive effort and method of compaction. They are nevertheless useful as a guide to specify, monitor and control field compaction.

FIELD DENSITY TESTS

Field density tests are used to measure the in-situ density of natural soils and compacted earthworks. The most common test, often used as a standard against which others are judged, is the sand replacement method. A hole is dug in the surface to the required depth. The soil removed is weighed and its moisture content taken. The hole is then filled with sand, carefully poured to achieve a known density, and the weight of sand used is recorded: this allows the volume of the hole to be found, hence the bulk and dry densities of the layer. To obtain good results it is important to carry out the test carefully. The hole should be cylindrical in shape and all the soil removed must be collected for weighing. To help achieve this, a metal tray with a hole cut in the centre is placed on the levelled surface and the hole dug through it so it can be used as a template and to collect the soil. Sand is poured from a special cylinder or bottle which delivers it through an egg-timer-like device to produce a constant density. Typical equipment is shown in Figure 4.9. The sand must be reasonably single-sized

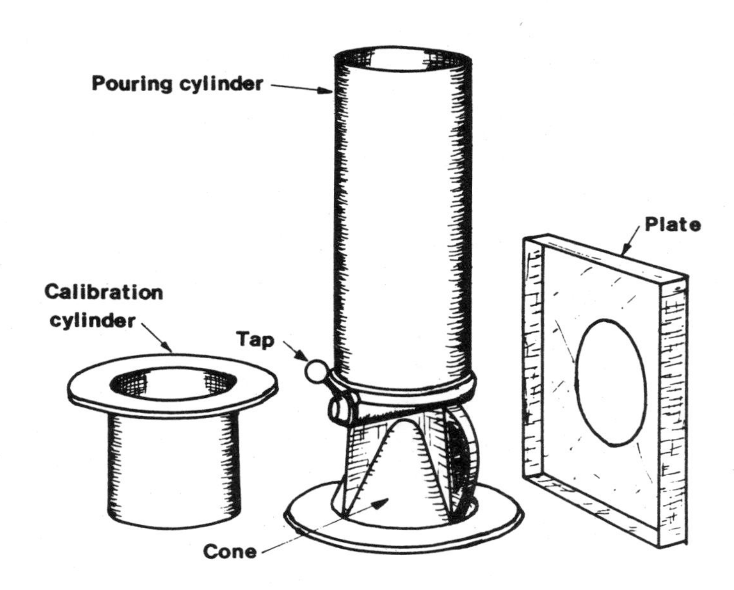

Figure 4.9 Field density test equipment using the sand replacement method

to avoid segregation and maintain constant density. Fairly accurate results can be obtained with this method but substantial errors can arise if the test is not carried out carefully.

The water replacement (or balloon density) method is similar to the sand replacement method but water is used instead of sand to measure the volume of the hole. The water is contained in a cylinder with a flexible rubber bag attached to the base, as shown in

Figure 4.10. This is placed over the hole and the cylinder is pressurised by means of a hand pump. This forces the water down into the rubber bag and inflates it until it fills the hole. The volume of water needed can be read directly from calibrations on the cylinder. The system is quicker than the sand density method and is popular in the U.S. and many other countries. However, it has generally been regarded with some scepticism in Britain out of fear that the bag may not completely fill the hole, so giving too high a value of density. It is not satisfactory in sands or soft clays which may distort owing to the pressure of the bag against the sides of the hole.

A common method used during ground investigation is to use a core cutter, typically 150mm or 200mm in diameter and about 200mm long. This is driven into the

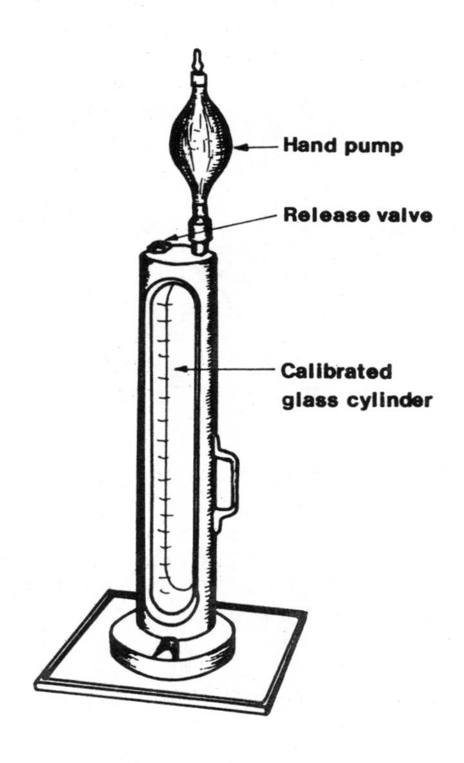

Figure 4.10 Balloon density apparatus

ground until soil protrudes through the top. It is then dug out and trimmed flush each end. It can then be weighed and, since its dimensions are known, the density of the soil can be calculated. Reasonably accurate results can usually be obtained in soft to stiff clays. Hard clays tend to fracture during driving and cannot be trimmed easily, whilst sands may compress during driving and often fall out of the cutter on excavation.

For those who like gadgets and do not mind the "black box" approach, the nuclear density meter offers a high-tech alternative to the more traditional methods. Bulk density is measured using a radio-isotope source and a gamma ray detector. The meter is placed on the surface and the detector records the amount of radiation which passes through the soil from a probe, attached to the meter, which is inserted into the ground: this is the direct transmission mode of operation. Alternatively, the meter can record radiation which has been reflected back by the soil from a source in the base of the meter: this is the backscatter mode of operation. From the radiation count over a fixed time period (typically 60 seconds), the bulk density can be determined. In addition, a radiation source is used to produce fast neutrons which collide with hydrogen atoms in the soil to produce slow neutrons. A slow neutron detector can be included in the equipment to measure the amount of hydrogen in the soil and, since nearly all hydrogen in soil is in the form of water, the moisture content can be obtained. Modern meters give reliable results with most soil types and are fairly quick to set up. Microprocessors convert readings directly to density and moisture content values which are displayed on a digital readout. The radiation source is small but it can still constitute a radiation hazard. Users should be properly trained and should wear radiation exposure badges to indicate whether they have been exposed to undue amounts of radiation.

THE CALIFORNIA BEARING RATIO TEST

Usually known as the CBR test, it is used in the design of pavements for roads and runways. A cylindrical plunger is pushed at a specific rate into a sample of soil and the force required at standard penetrations (typically 2.5mm and 5.0mm) is recorded. Values obtained are expressed as a percentage of standard values to give the CBR value of the soil. The test does not measure any fundamental property of the soil and is often scorned by those who prefer to keep their soil mechanics on a more theoretical plane but it is specified by agencies in many parts of the world as a basis for pavement design. Ironically, it is no longer used in California, its original home.

It is usually performed on samples made up in the laboratory in a special mould but it may be carried out as a field test, using equipment mounted on the back of a vehicle. A typical test set-up is shown in Figure 4.11. Results are strongly influenced by the method of sample and preparation and test details: its density and moisture content; whether the sample is soaked before testing; whether and how many surcharge weights are placed on the sample during the test; whether both ends of the sample are tested or just the base; even what method of compaction is used. The details of the method used must be consistent with what has been assumed in the particular pavement design method to be employed. Failure to comply with some quite small details may result in significant errors in the test results.

CBR tests require about 7kg of soil and individual results can show quite a large scatter, especially with weak clay soils, so a number of tests should be performed before a design value is selected. In addition, a compaction test must first be performed, so that the sample can be compacted to a suitable condition. This can seem something of an overkill when a few estate roads are being constructed. Fortunately, the Transport and Road Research Laboratory has published a correlation between CBR and plasticity index for a variety of British soils. This enables a CBR-based design to be used without the drudgery of actually having to carry out CBR tests.

PERMEABILITY TESTS

Laboratory permeability tests are useful where flow

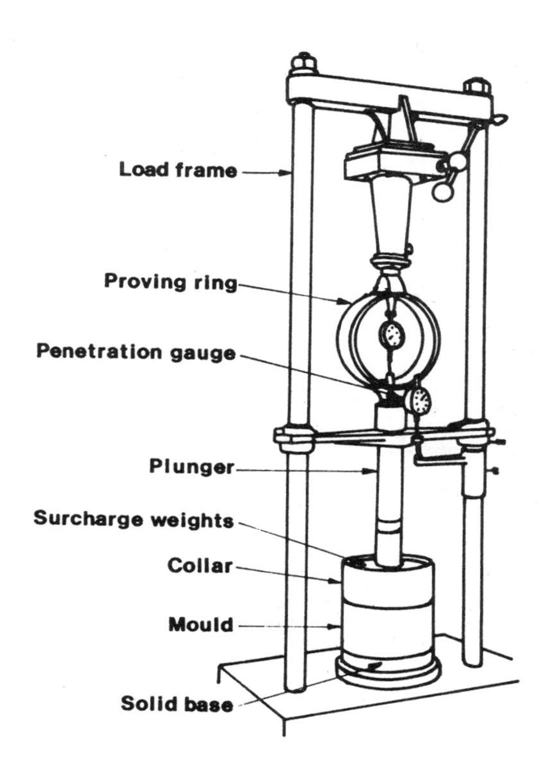

Figure 4.11 CBR test equipment

problems are encountered - to estimate the quantity of flow through a dam or the required pumping capacity from an excavation for instance - but are not included in the majority of testing programs. In essence, the tests consist simply of applying a head of water to one end of a cylindrical specimen of soil and measuring the flow through it. There are two basic types of test: constant head and falling head as shown diagrammatically in Figure 4.12. The constant head test allows relatively large quantities of water to flow through the sample, and to be measured, and is suitable for the more permeable soils (sands and gravels). The falling head test, which allows more

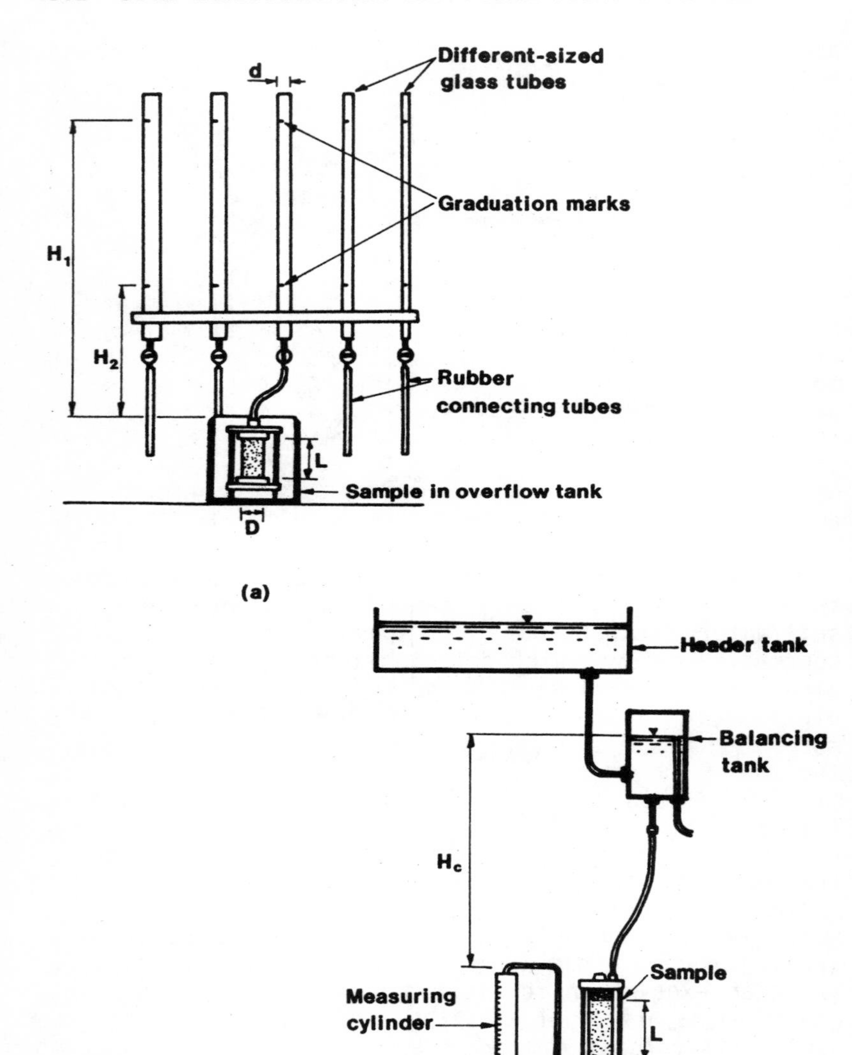

Figure 4.12 Basic layout of permeability test equipment: (a) falling head test; (b) constant head test

accurate measurement of small quantities of water, is suitable for less permeable soils (silts and clays). Specimens may be made from undisturbed or remoulded soil samples. Tests are sometimes carried out in conjunction with field permeability tests.

CHEMICAL TESTS

The chemistry of the soil is usually of interest to the engineer only to the extent that it might adversely affect his structure. Buried structures, which are most likely to be made of steel, concrete and clay bricks, are adversely affected by acid conditions, by sulphates and, occasionally, by chlorides. The presence of organic matter is also undesirable because of the acid conditions which may be generated by decaying material.

The acidity or pH test is the simplest of the chemical tests to carry out. It can be used to measure the acidity of either a sample of groundwater or a soil/water suspension. Laboratories usually use special pH meters which detect slight variations in the electrical resistivity of the water. An electrode is dipped into a sample and a reading obtained. For the more casual user a simple chemical test is available. The water sample is mixed with barium sulphate and a mixture of indicator solutions is added. The colour of the resulting cocktail is compared with a standard colour chart to give the pH value. Soil pH kits, using this method, are commercially available.

Sulphate content tests may also be carried out on both soil and groundwater samples. Not all sulphates are soluble in water and, when soil samples are to be tested, the specification should state whether the water soluble sulphate content or the total (acid soluble) sulphate content is required. Normally, the total sulphate content test is used but where results are unacceptably high the water soluble sulphate content is checked. If this is low, the designer has to consider whether the groundwater conditions will be acid enough to mobilise the insoluble sulphates over a period of time. He must then decide whether to base any protection against sulphate attack on the total or the acid soluble sulphate content.

The organic matter content of a soil may be determined by burning an oven-dried sample over a gas flame and noting the loss in weight. This simple method gives an approximate indication of the organic matter content but results may be unreliable because many soils lose weight when heated owing to a breakdown of inorganic constituents. For instance, water may be driven off from the adsorption complex of clay minerals or from sulphates, or carbon dioxide may be lost from carbonates. An alternative method is to chemically oxidise the organic carbon using potassium dichromate solution and concentrated sulphuric acid. Even this test, which relies on certain assumptions about the proportion of carbon in the organic matter and what proportion of this is oxidised, should not be expected to produce highly accurate results. Fortunately it is not normally necessary to know the organic matter content of a soil to a high degree of accuracy in order to make an engineering judgement.

Many of the chemical tests on soils, including sulphate and chloride determinations, are complex and difficult to carry out; test details have not been included here since it would serve little purpose. The equipment and skills needed to perform the tests differ greatly from those normally associated with a soils laboratory and many small laboratories prefer to contract out this work to suitably equipped chemical laboratories.

Chapter 5

THE GEOTECHNICAL REPORT

CONTENTS AND LAYOUT OF A TYPICAL REPORT

As discussed in Chapter 1, reports fall broadly into two categories: the basic, "factual", site investigation report which simply records the findings of the site investigation; and the "comprehensive" or "engineering" report which, in addition, interprets those findings and makes specific recommendations. Most organisations producing geotechnical reports like to adopt a particular "house" style so that all their reports have a similar layout. This is usually fairly easy to achieve except for the occasional complex or unusual project for which there may be good reason to change the usual order. The layout may be specified by the client but this is unusual except for a few large organisations such as some government departments.

There will obviously be variations in the layout and presentation of reports from different sources but most reports follow a similar overall pattern.

A factual report typically contains the items described below.

Introduction

The project is briefly described and the scope of the report defined. The name of the client and his consulting engineers are given. Location of the site is often included in this section.

Topography and geology

The site, and possibly the surrounding area, are briefly described. This will include general descriptions (whether the area is flat, has rolling hills or whatever); will mention specific features (such as a stream or pond); and will state what is on the site (possibly trees or buildings). A brief summary of published geological information is normally also given in this section.

Site work

Details of the extent of site work are given, including the number of trial pits, boreholes and probes and the methods used. The types and numbers of any field tests are also included. Overall dates of site working are given.

Laboratory testing

The types and numbers of laboratory tests are given, with overall dates for the testing program.

Summary of ground conditions

It is usual, even in a factual report, to summarise what conditions were encountered during the site work.

Test results

Results of all field and laboratory tests are presented in the form of tables and graphs.

Trial pit and borehole records

Details of each trial pit and borehole are given diagrammatically, as described later in this chapter.

Site plan

A plan or plans showing borehole and trial pit locations.

Notes

At some point in the report there is usually a set of notes giving standard abbreviations and symbols used. Details of some of the standard tests may be explained and references may be made to the standards used in carrying out the work. There is usually a general disclaimer to protect the site investigation contractor should problems arise later due to ground conditions which were not revealed during the ground investigation.

Engineering reports usually contain all the items given in the factual report, either as part of the report itself or as an appendix, plus the additional items outlined below.

Description of the project

This is usually discussed more fully than in the factual report. It will include, where possible, such items as the size and type of individual buildings and any particular considerations such as permissible settlements.

Results of the desk study

The desk study carried out by site investigation firms for a factual report is usually fairly rudimentary and may only consist of checking the appropriate geological maps. For the engineering report the desk study section may be substantial but could be virtually non-existent, depending on how much information is available.

Discussion of ground conditions

Again, this is much more fully covered than the simple summary which usually appears in the factual report. Where appropriate, it will refer to possible seismic, hydrological or mining problems, or any other special factors.

Soil properties for design

Design values of soil properties, such as shear strength and compressibility, are given for each of the foundation soils encountered. Variations in properties within a given soil, both across the site and with depth, should be indicated. The extent and location of each of the soil types and their distinguishing features should also be included.

Foundation design

Foundation types are recommended, along with suggested founding depths, sizes, permissible loadings, expected settlement and any other relevant details.

Calculations

The recommended values of soil properties and foundation design details given in the previous two sections should be supported by clearly laid out calculations. The assumptions made and the design methods used should be stated so the work can be checked by others and values can easily be amended if new information becomes available.

Site plans

These are usually more detailed than for the factual report and normally show the proposed project layout.

Other items

These may include such items as a brief for the resident engineer; a warning that specific problems may arise which will require special treatment or a redesign of certain features; or a recommendation that further investigation be carried out.

The above description refers to typical items which would be expected in a well-prepared engineering report for a substantial project. Some of these items would require only brief treatment for small, simple projects but another reason for glossing over many of these points, it must be admitted, is that not all reports measure up to a standard that could be described as well-prepared. In this respect, it is useful to know what should have been included in a report and to be on guard against possible errors and omissions.

Not only do the layouts of reports follow a broadly similar pattern but the information provided in each section is usually given using standardised terms and presentation. This helps to convey information effectively and reduces the risk of ambiguity or misunderstanding, providing it is not taken to pedantic extremes. The ways in which information is presented and the terms used are described in the following sections of this chapter.

SOIL DESCRIPTIONS

For civil engineering purposes, soil may be taken to include all natural deposits which can be separated by gentle mechanical means and excavated without blasting. This description will include many materials which are classified geologically as rocks, and geologists often refer to soil defined in this general way as "regolith". As rocks slowly weather into soils and soils are gradually transformed into rocks, so the distiction between the two often becomes blurred. Most soils fall into one of two major groups: transported and residual. Transported soils predominate in temperate latitudes and residual soils are more common in tropical regions. Other minor groups of soils include organic, volcanic and evaporite deposits. A broad classification of soils according to their geological origins is given in Table 5.1.

Soil descriptions are made from field observations and from inspection of disturbed and undisturbed samples taken from cuttings, excavations and boreholes. For engineering purposes, descriptions are usually based on particle size and plasticity properties. They

TABLE 5.1
BROAD CLASSIFICATION OF SOILS ACCORDING TO GEOLOGICAL ORIGIN

CLASSIFICATION AND PROCESS OF FORMATION	NATURE OF DEPOSITS
Residual Chemical weathering of parent rock with little or no movement of particles.	Product of complete weathering is a clay whose type depends mainly on the weathering process. Products of partial weathering are more stoney and depend more on rock type. Soil becomes more compact, more stoney and less weathered with increasing depth.
Alluvial Materials transported and deposited by water action.	Vary from finest clays to very coarse gravel and boulders. Soils usually show pronounced stratification. River gravels are usually rounded.
Colluvial Materials transported by gravity.	Includes screes, avalanches, landslips, hillside creep, downwash material, and solifluxion deposits. Varies from clays to boulders. Material is usually heterogeneous with a wide range of particle sizes. Often termed hillwash or head deposits.
Glacial Materials transported and deposited by glacial ice or by melt waters from glaciers.	Glacial till and morain deposits usually have broad gradings ranging from clay to boulders. Grain size in the outwash material decreases with distance from the source of melt water. Stratification in morains and till is usually heterogeneous but outwash deposits give rise to laminated (varved) silt and clay in glacial lakes. Grains are typically angular.
Aeolian or loessial Materials transported and deposited by wind.	Highly uniform gradation with indistinct or no stratification. Typically silt or fine sand sized but sometimes the surface is covered by a single layer of pebbles. Loess typically has a secondary structure of vertical cracks, joints and root holes.
Organic Formed in place by growth and decay of plants.	Peats are dark coloured, fibrous or amorphous and highly compressible. Mixtures of fine sediment and organic matter produce organic silts and clays.
Volcanic Ash and pumice deposited in volcanic eruptions.	Silt sized particles along with larger volcanic debris. Particles are highly angular and often vesicular. Weathering produces a highly plastic, sometimes expansive clay. The weathered consolidated deposits sometimes form a light, easily-worked stone.
Evaporites Materials precipitated or evaporated from solutions of high salts content.	Forms cemented soils or soft sedimentary rocks. Includes oolites precipitated from calcium in sea water and gypsum precipitated from sulphate-rich playa lakes in deserts. Evaporites may form as a hard crust just below the surface in arid regions.

should contain information on some or all of the following properties:

(1) field strength or compactness,
(2) structure (bedding, discontinuities etc.) and state of weathering,
(3) colour,
(4) particle shape and composition,
(5) soil name, based on particle size,
(6) reference to inclusions of other materials.

Properties are usually described in approximately the order given above. The directions and examples given in Table 5.2 summarise the information which should be included and show the order and style of descriptions that are usually used. Some of the properties included in the descriptions are discussed in more detail below.

Compactness and relative density

The term "compactness" is used to describe the extent to which granular soils are compacted together in their natural state. It is defined in terms of relative density, D; the proportion of voids in the soil compared with the proportions corresponding to the loosest and densest possible states of packing of that soil. It can also be defined in terms of dry densities as shown in the formal definition below.

$$D = \frac{e_{max} - e}{e_{max} - e_{min}} = \frac{\gamma_{max}}{\gamma} \cdot \frac{\gamma - \gamma_{min}}{\gamma_{max} - \gamma_{min}}$$

where e, e_{max} and e_{min} are the voids ratios (volume of voids as a proportion of volume of solids) for the soil in its natural state and at its loosest and densest states of compaction, respectively, and γ, γ_{min} and γ_{max} are the corresponding dry densities.

Thus, relative density is a number between 0 (for soil in its loosest state) and 1 (for soil in its densest state), or a percentage between 0 and 100. Descriptive terms are given to the various properties of relative density as indicated in Table 5.3.

The concept of compactness or relative density is a useful one for granular soils because it can be correlated with a number of properties related to their strength and deformation characteristics; and against

TABLE 5.2 SOIL DESCRIPTIONS

DESCRIBING A SOIL

Descriptions should include some or all of the items listed in the headings below. By working from left to right across the columns below and noting the properties of a soil sample, a useful description of the soil in a standardised format will be obtained.

EXAMPLES

Loose brown subangular very sandy fine to coarse GRAVEL with small pockets of soft grey clay.
Soft laminated dark blue silty CLAY.
FILL (stiff orange-brown clay with scattered brick fragments).

Shear strength/relative density			Structure	Colour	Particle shape/ composition etc	Type of particle			Inclusions
CLAYS	c(kN/m2)	CHARACTERISTICS	Intact	Grey	Angular Subangular	**PARTICLE**		**SIZE(mm)**	..with shells.
Very soft	< 20	Exudes between fingers when squeezed.	Fissured	Brown	Subrounded Rounded	Clay		0.002	..with scattered cobbles and boulders.
Soft	20–40	Moulded by light finger pressure.	Stratified	Blue-grey	Flat Elongated		fine	0.002–0.006	
Firm	40–75	Moulded by strong finger pressure.	Laminated	Mottled yellow and brown	Irregular	Silt	medium	0.006–0.02	..with layers or lenses of fine sand.
Stiff	75–150	Can be indented by thumb.	Heterogeneous		Rough Smooth		coarse	0.02–0.06	
Very stiff	150–300	Can be indented by thumb nail.	Fibrous	Dark Green	Polished		fine	0.06–0.2	..with some shell fragments.
Hard	> 300			Yellowish	Sandstone Limestone	Sand	medium	0.2–0.6	etc.
SANDS	SPT*		etc.	etc.	Granite		coarse	0.6–2.0	
Very loose	< 4				Highly plastic Non-plastic		fine	2–6	
Loose	4–10	Can be dug by spade. 50mm peg easily driven.				Gravel	medium	6–20	
Medium dense	10–30				etc.		coarse	20–60	
Dense	30–50	Needs pick for excavation. 50mm peg hard to drive.							
Very dense	> 50					Cobbles		60–200	
						Boulders		200	

*Standard penetration test N-values.

TABLE 5.3
DESCRIPTIVE TERMS FOR RELATIVE DENSITY, WITH EQUIVALENT SPT N-VALUES

Relative density (%)	Descriptive term	N-values
0-15	Very loose	0-4
15-35	Loose	4-10
35-65	Medium dense	10-30
65-85	Dense	30-50
85-100	Very dense	50+

the soil's resistance to penetration, as measured by a standard penetration test (SPT) or a cone penetration test for instance. This means that various important properties of granular soils can be inferred from the results of some fairly simple in-situ tests, which conveniently gets around the near-impossibility of obtaining undisturbed samples and testing them. Table 5.3 also shows equivalent SPT N-values corresponding to various proportions of relative density.

For simple trial pit work, where penetration testing is unlikely to be carried out, compactness can be roughly assessed by digging into the deposit, using the simple rules indicated in column 1 of Table 5.2. If the deposit can be dug easily using a spade or shovel and if a 50mm square peg can be easily driven into it, then it is described as "loose". If a pick is needed to loosen the material and the peg is hard to drive then it is described as "dense" or "compact". The dividing line corresponds to a relative density of about 35%, or an N-value of 10. This rough-and-ready method is sufficient for many purposes but it is important that the designer knows the basis of relative density descriptions so he can decide how conservative an approach to take, and this should be clearly stated in the geotechnical report.

Shear strengths

For most clays the undrained shear strength is independent of the confining pressure when it is sheared in a quick undrained triaxial test. Similarly, when rapidly sheared in a shear box, it is independent of normal stress on the specimen. The shear strength can therefore be defined by a single stress value. Since the quick undrained shear strength is used in foundation design, it is obviously an important property. In addition, settlement is more of a problem where clays with low shear strength are encountered. Because of its importance, shear strength is included in soil descriptions wherever possible. As can be seen in Table 5.2, there are well-defined descriptive terms for strength which relate to precise ranges of shear strength. For instance, where shear strength measurements of a clay fall within the range 45-60kN/m^2, it would be described as "firm". If shear strengths were in the range 60-90kN/m^2, it would be described as "firm to stiff". Shear strength values may be based on laboratory tests, such as the triaxial, shear box or unconfined compression tests; on pocket penetrometer tests; or on resistance to moulding in the hand, using the simple tests described in the table. This should be specified in the report, because the accuracy and reliability of the strength descriptions will depend on the method of testing used. Since pocket penetrometers are so cheap and convenient to use, there is little excuse nowadays for basing strength descriptions on hand moulding tests.

Structure

A description of structure is useful when trying to assess whether the clay found in different boreholes is the same basic material or an altogether different deposit. The structure can also affect the engineering properties of a soil and an accurate description may be important to make a proper judgement. For instance, fissured or laminated soils are likely to consolidate much more quickly than would be expected from the results of oedometer tests. Also, a fissured clay is likely to be considerably weaker as a soil mass, than small-scale testing would suggest. Failure to notice fissuring or lamination, or to include it in the description, could lead to inappropriate or even unsafe designs.

Colour

Like structure, colour is useful to identify the extent of a soil deposit. Standard colour descriptions are available but are rarely used, and consistency in description is usually quite sufficient. Although colour usually helps to identify and distinguish between soils, cases may arise where the same soil possesses different colours on the same site. For instance, a soil might be brown above the water table and blue below it.

Particle shape/composition/plasticity and type of particle

As described in Chapter 4, particle size and shape are important for coarse-grained soils whereas the plasticity properties are important for fine-grained soils. Most soils are not simply sands or gravels but contain a range of particle types.

To describe the composition of particles in a soil, terms such as "sandy GRAVEL" or "clayey silty fine to medium SAND" are used. Note that the main soil type is given in capitals and that commas are not used. Some examples of appropriate descriptions for sand-gravel and clay-gravel mixtures are given in Table 5.4.

The classification of a soil into a silt or sand, for instance, depends on the size of the soil particles, as indicated in Table 5.2. However, descriptions are often based on visual inspection rather than on laboratory test results.

Descriptions of the plasticity of clays are often omitted. When they are included, terms such as "highly plastic" are used, based on the Unified or British Standard classification systems. These are described later.

Inclusions

This can include any particles found within the soil mass such as shells, bands of sand, clay lumps, coal fragments or broken pottery.

TABLE 5.4
DESCRIPTIONS FOR COMPOSITE SOILS

Description	Composition
Slightly sandy GRAVEL (GRAVEL with a little sand)	<5% sand
Sandy GRAVEL (GRAVEL with some sand)	5-20% sand
Very sandy GRAVEL	>20% sand
SAND/GRAVEL	about equal
Very gravelly SAND	>20% gravel
Gravelly SAND	5-20% gravel
Slightly gravelly SAND (SAND with scattered gravel)	<5% gravel
CLAY with scattered gravel	<5% gravel
CLAY with some gravel	5-35% gravel
Gravelly CLAY	35-65% gravel

TABLE 5.5
TYPICAL COMPRESSIBILITY VALUES FOR SOILS AND DESCRIPTIVE TERMS USED

Descriptive term	Coefficient of volume compressibility, m_v		Examples of clay types
	(m^2/MN)	(ft^2/ton)	
Very low compressibility	<0.05	<0.005	Heavy over-consolidated boulder clays, stiff weathered rocks (e.g. weathered mudstone) and hard clays.
Low compressibility	0.05-0.1	0.005-0.01	Boulder clays, marls, very stiff clays and stiff tropical red clays.
Medium compressibility	0.1-0.3	0.01-0.03	Firm clays, glacial outwash clays, lake deposits, weathered marls, firm boulder clays, normally consolidated clays at depth and firm tropical red clays.
High compressibility	0.3-1.5	0.03-0.15	Normally consolidated alluvial clays such as estuarine and delta deposits, and sensitive clays.
Very high compressibility	>1.5	>0.15	Highly organic alluvial clays and peats.

Compressibility

Compressibility descriptions are not usually included in soil descriptions but, when they are, they are based on the values of volume compressibility, m_v, as given in Table 5.5, which also gives typical ranges for various clay types.

Weathering grades

The degree of weathering of residual soils and weathered rocks strongly influences their engineering properties. It is expressed in terms of weathering grades incorporated into the soil descriptions. Although particularly appropriate for tropical residual soils, weathering grades can also be used advantageously to describe deposits such as weathered chalk and marl.

Residual soils are formed by the chemical decomposition of rocks with little or no transportation of particles. With these soils plasticity and grading vary according to the pretreatment of the material. In many cases, soils form a continuous, cemented mass which breaks down on excavation and continues to break down further on handling. For such material, the concept of grading, when applied to the in-situ state, has no meaning. There is often no distinct boundary between a soil and its parent rock and it is convenient to classify residual soils in terms of the degree of weathering that has taken place.

A secondary effect of chemical weathering, which occurs in areas with pronounced dry periods, is the precipitation of dissolved matter which has been leached out from the soil during wetter parts of the year. The effects of repeated cycles of wet and dry periods is to build up a surface of concretionary material (laterite, caliche, etc.) a little way below the surface. Although this is not strictly a weathering condition, it is convenient to include it in the classification.

Descriptions of the material usually follow the same basic pattern as those given in Table 5.2, but modified to take into account the grading limitations described above. This is accompanied by the weathering grade, as given in the classification shown in Table 5.6. For instance:

(1) Stiff, faintly laminated, grey silty clay with scattered gravel-size rock fragments: WEATHERED ROCK grade IV-V.

(2) WEATHERED ROCK grade III (yellow-brown mottled gneiss with red sandy clay and occasional bands of fractured quartz).

TABLE 5.6
IDEALISED WEATHERING PROFILE AND DESCRIPTIVE TERMS USED FOR WEATHERED ROCKS AND RESIDUAL SOILS

WEATHERING GRADE	DESCRIPTION
VII CONCRETION	Hard concretionary deposit, typically 0.5-1.5m thick. Usually needs ripping. Good roadmaking material.
VI SOIL	No recognisable rock texture. May occur above and below any concretions. Surface layer contains humus and plant roots.
V COMPLETELY WEATHERED	Rock completely decomposed by weathering in place but texture still recognisable. Can be excavated by hand. Unsuitable for foundations of large structures.
IV HIGHLY WEATHERED (< 50% rock)	Rock so weakened by weathering that fairly large pieces can be broken and crumbled in the hands. Sometimes recovered as core in careful rotary drilling. Erratic pressure of boulders makes unreliable foundation.
III MODERATELY WEATHERED (50% - 90% rock)	Considerably weathered throughout. Possessing some strength - large pieces cannot be broken by hand. Reasonable core recovery. Often limonite stained. Difficult to rip. Fairly good foundation material and may be suitable for pavement construction.
II SLIGHTLY WEATHERED (< 90% rock)	Distinctly weathered through much of the rock fabric with slight limonite staining. Strength approaches that of fresh rock. Requires explosives for excavation. Highly permeable through open joints.
I FRESH ROCK	Fresh rock may have some limonite stained joints, indicating water percolation.

IN MANY PROFILES, GRADES MAY BE ABSENT OR IN A DIFFERENT ORDER.

SOIL CLASSIFICATION SYSTEMS

The aim of most soil classification systems is to enable an assessment to be made of the engineering properties of soils using visual inspection and simple tests. Used in conjunction with empirical design methods, classification systems set down in a formal manner the accumulated international experience of many individuals, allowing the engineer to benefit from a far wider range of practical knowledge than he could gain from personal experience alone. Also, soil classification aids the engineer in the interpretation of soil strength test results by helping him to identify the separate soil types occurring at a site and the extent of each type.

There are many soil classification systems but probably the most common is the Unified System, or variations of it, originally proposed by Arthur Casagrande in 1948. The system has been adopted by the American Society for Testing and Materials and is published as ASTM standard D2847. Soil is divided into classes, based on grading and plasticity tests, as indicated in Table 5.7 and Figure 5.1. A variation of this system, used in Britain, is the BS classification system which is described in detail in British Standard 5930:1981. Names and descriptive terms used in the BS classification are given in Table 5.8 and the method of classifying soils is given in Table 5.9 and Figure 5.2.

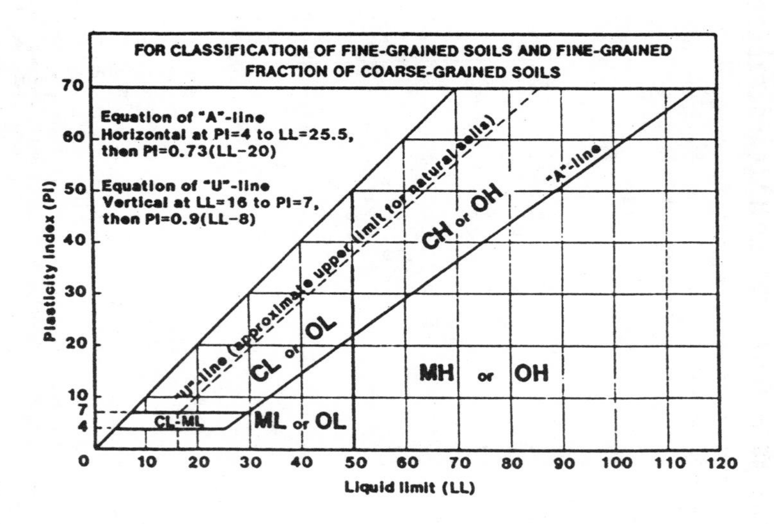

Figure 5.1 Casagrande plasticity chart as used in the unified (ASTM) soil classification system

TABLE 5.7
THE ASTM (UNIFIED) SOIL CLASSIFICATION SYSTEM (AFTER ASTM D2847-85)

Criteria for Assigning Group Symbols and Group Names Using Laboratory Tests[1]				Soil Classification	
				Group Symbol	Group Name[2]
Coarse-Grained Soil More than 50% retained on No. 200 (0.075mm) sieve	Gravels More than 50% of coarse fraction retained on No. 4 (4.75mm) sieve	Clean Gravels Less than 5% fines[3]	Cu ≤ 4 and 1 ≤ Cc ≤ 3[5]	GW	Well-graded gravel[6]
			Cu ≤ 4 and/or 1 > Cc > 3[5]	GP	Poorly graded gravel[6]
		Gravels with Fines More than 12% fines[3]	Fines classify as ML or MH	GM	Silty gravel[6 7 8]
			Fines classify as CL or CH	GC	Clayey gravel[6 7 8]
	Sands 50% or more of coarse fraction passes No. 4 (4.75mm) sieve	Clean Sands Less than 5% fines[4]	Cu ≥ 6 and 1 ≤ Cc ≤ 3[5]	SW	Well-graded sand[9]
			Cu < 6 and/or 1 > Cc > 3[5]	SP	Poorly graded sand[9]
		Sands with Fines More than 12% fines[4]	Fines classify as ML or MH	SM	Silty sand[7 8 9]
			Fines classify as CL or CH	SC	Clayey sand[7 8 9]
Fine-Grained Soils 50% or more passes the No. 200 sieve	Silts and Clays Liquid limit less than 50	inorganic	PI > 7 and plots on or above "A" line[10]	CL	Lean clay[11 12 13]
			PI < 4 or plots below "A" line[10]	ML	Silt[11 12 13]
		organic	Liquid limit - oven dried / Liquid limit - not dried < 0.75	OL	Organic clay[11 12 13 14]
					Organic silt[11 12 13 15]
	Silts and Clays Liquid limit 50 or more	inorganic	PI plots on or above "A" line	CH	Fat clay[11 12 13]
			PI plots below "A" line	MH	Elastic silt[11 12 13]
		organic	Liquid limit - oven dried / Liquid limit - not dried < 0.75	OH	Organic clay[11 12 13 16]
					Organic silt[11 12 13 17]
Highly organic soils	Primarily organic matter, dark in colour, and organic odour			PT	Peat

1. Based on the material passing the 3-in. (75 mm) sieve.
2. If field sample contained cobbles or boulders, or both, add "with cobbles or boulders, or both" to group name.
3. Gravels with 5 to 12% fines require dual symbols:
   GW-GM well-graded gravel with silt
   GW-GC well-graded gravel with clay
   GP-GM poorly graded gravel with silt
   GP-GC poorly graded gravel with clay
4. Sands with 5 to 12% fines require dual symbols:
   SW-SM well-graded sand with silt
   SW-SC well-graded sand with clay
   SP-SM poorly graded sand with silt
   SP-SC poorly graded sand with clay
5. $Cu = D_{60}/D_{10}$ $Cc = \frac{(D_{30})^2}{D_{10} \times D_{60}}$
6. If soil contains 15% sand, add "with sand" to group name.
7. If fines classify as CL-ML, use dual symbol GC-GM, or SC-SM.
8. If fines are organic, add "with organic fines" to group name.
9. If soil contains 15% gravel, add "with gravel" to group name.
10. If Atterberg limits plot in hatched area, soil is a CL-ML, silty clay.
11. If soil contains 15 to 29% plus No. 200, add "with sand" or with gravel", whichever is predominant.
12. If soil contains 30% plus No. 200, predominantly sand, add "sandy" to group name.
13. If soil contains 30% plus No. 200, predominantly gravel, add "gravelly" to group name.
14. PI 4 and plots on or above "A" line.
15. PI 4 or plots below "A" line.
16. PI plots on or above "A" line.
17. PI plots below "A" line.

ROCK DESCRIPTIONS

Like soil descriptions, rock descriptions follow an established pattern. Descriptions used for engineering purposes are different from those used for geological work. Broad classifications of rock type are usually sufficient - detailed geological names are not necessary - but details of the rock quality, strength and structural features are important. Two indicators of rock quality are often used. These are "recovery" and "rock quality designation" (RQD):

$$\text{Recovery (\%)} = \frac{\text{length of core recovered}}{\text{length of core run}} \times 100$$

$$\text{RQD (\%)} = \frac{\text{length recovered in sound lengths of 100mm or more}}{\text{length of core run}} \times 100$$

These give an indication of the strength of the rock mass but they can be used only as a rough guide because values obtained depend on the diameter of the core, the method of drilling and the skill of the driller. Describing and evaluating rocks for engineering purposes requires specialist knowledge and this work is usually undertaken by engineering geologists.

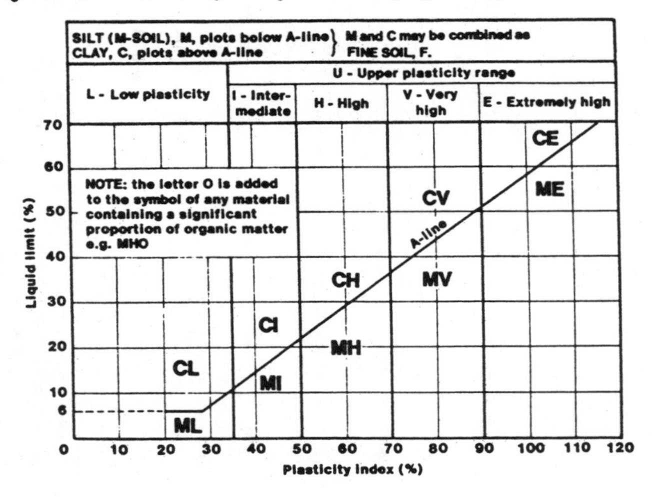

Figure 5.2 Casagrande plasticity chart as used in the British Standard soil classification system.

TABLE 5.8
NAMES AND DESCRIPTIVE TERMS USED IN THE BRITISH STANDARD SOIL CLASSIFICATION SYSTEM (BS5930:1981)

		Descriptive name	Letter
Coarse components	Main terms	GRAVEL	G
		SAND	S
	Qualifying terms	Well graded	W
		Poorly graded	P
		Uniform	Pu
		Gap graded	Pg
Fine components	Main terms	FINE SOIL, FINES may be differentiated into M or C	F
		SILT (M-SOIL)* plots below A-line of plasticity chart of figure 31 (of restricted plastic range)	M
		CLAY plots above A-line (fully plastic)	C
	Qualifying terms	Of low plasticity	L
		Of intermediate plasticity	I
		Of high plasticity	H
		Of very high plasticity	V
		Of extremely high plasticity	E
		Of upper plasticity range† incorporating groups I, H, V and E	U
Organic components	Main term	PEAT	Pt
	Qualifying term	Organic may be suffixed to any group	O

*See note 5 following table 8.

†This term is a useful guide when it is not possible or not required to designate the range of liquid limit more closely, e.g. during the rapid assessment of soils.

BOREHOLE AND TRIAL PIT LOGS

It is essential that the information gained from the ground investigation is fully documented in a well-presented form. For each borehole and trial pit a formal log is kept, giving details of the size of hole and method of excavation or boring; dates work was carried out; the soils encountered with depths; the types and depths of samples taken or in-situ tests carried out; groundwater conditions encountered; any piezometers or other instrumentation installed; and any other special condition or problems encountered during boring or excavation. It is usual to present this

TABLE 5.9
THE BRITISH STANDARD SOIL CLASSIFICATION SYSTEM FOR ENGINEERING PURPOSES (BS 5930:1981)

Soil groups (see note 1)			Subgroups and laboratory identification				
GRAVEL and SAND may be qualified Sandy GRAVEL and Gravelly SAND, etc. where appropriate (See 41.3.2.2)			Group symbol (see notes 2 & 3)	Subgroup symbol (see note 2)	Fines (% less than 0.06 mm)	Liquid limit %	Name
COARSE SOILS less than 35 % of the material is finer than 0.06 mm	GRAVELS More than 50 % of coarse material is of gravel size (coarser than 2 mm)	Slightly silty or clayey GRAVEL	G: GW	GW	0 to 5		Well graded GRAVEL
			G: GP	GPu GPg			Poorly graded/Uniform/Gap graded GRAVEL
		Silty GRAVEL	G-F: G-M	GWM GPM	5 to 15		Well graded/Poorly graded silty GRAVEL
		Clayey GRAVEL	G-F: G-C	GWC GPC			Well graded/Poorly graded clayey GRAVEL
		Very silty GRAVEL	GF: GM	GML, etc	15 to 35		Very silty GRAVEL; subdivide as for GC
		Very clayey GRAVEL	GF: GC	GCL GCI GCH GCV GCE			Very clayey GRAVEL (clay of low, intermediate, high, very high, extremely high plasticity)
	SANDS More than 50 % of coarse material is of sand size (finer than 2 mm)	Slightly silty or clayey SAND	S: SW	SW	0 to 5		Well graded SAND
			S: SP	SPu SPg			Poorly graded/Uniform/Gap graded SAND
		Silty SAND	S-F: S-M	SWM SPM	5 to 15		Well graded/Poorly graded silty SAND
		Clayey SAND	S-F: S-C	SWC SPC			Well graded/Poorly graded clayey SAND
		Very silty SAND	SF: SM	SML, etc	15 to 35		Very silty SAND; subdivided as for SC
		Very clayey SAND	SF: SC	SCL SCI SCH SCV SCE			Very clayey SAND (clay of low, intermediate, high, very high, extremely high plasticity)

TABLE 5.9 (CONTINUED)

FINE SOILS more than 35 % of the material is finer than 0.06 mm	Gravelly or sandy SILTS and CLAYS 35 % to 65 % fines	Gravelly SILT	FG MG	MLG, etc			Gravelly SILT; subdivide as for CG
		Gravelly CLAY (see note 4)	CG	CLG CIG CHG CVG CEG		< 35 35 to 50 50 to 70 70 to 90 > 90	Gravelly CLAY of low plasticity of intermediate plasticity of high plasticity of very high plasticity of extremely high plasticity
		Sandy SILT (see note 4)	FS MS	MLS, etc			Sandy SILT; subdivide as for CG
		Sandy CLAY	CS	CLS, etc			Sandy CLAY; subdivide as for CG
	SILTS and CLAYS 65 % to 100 % fines	SILT (M-SOIL)	F M	ML, etc			SILT; subdivide as for C
		CLAY (see notes 5 & 6)	C	CL CI CH CV CE		< 35 35 to 50 50 to 70 70 to 90 > 90	CLAY of low plasticity of intermediate plasticity of high plasticity of very high plasticity of extremely high plasticity
ORGANIC SOILS		Descriptive letter 'O' suffixed to any group or sub-group symbol.					Organic matter suspected to be a significant constituent. Example MHO: Organic SILT of high plasticity.
PEAT		Pt					Peat soils consist predominantly of plant remains which may be fibrous or amorphous.

NOTE 1. The name of the soil group should always be given when describing soils, supplemented, if required, by the group symbol, although for some additional applications (e.g. longitudinal sections) it may be convenient to use the group symbol alone.

NOTE 2. The group symbol or sub-group symbol should be placed in brackets if laboratory methods have not been used for identification, e.g. (GC).

NOTE 3. The designation FINE SOIL or FINES, F, may be used in place of SILT, M, or CLAY, C, when it is not possible or not required to distinguish between them.

NOTE 4. GRAVELLY if more than 50 % of coarse material is of gravel size. SANDY if more than 50 % of coarse material is of sand size.

NOTE 5. SILT (M-SOIL), M, is material plotting below the A-line, and has a restricted plastic range in relation to its liquid limit, and relatively low cohesion. Fine soils of this type include clean silt-sized materials and rock flour, micaceous and diatomaceous soils, pumice, and volcanic soils, and soils containing halloysite. The alternative term 'M-soil' avoids confusion with materials of predominantly silt size, which form only a part of the group.

Organic soils also usually plot below the A-line on the plasticity chart, when they are designated ORGANIC SILT, MO.

NOTE 6. CLAY, C, is material plotting above the A-line, and is fully plastic in relation to its liquid limit.

information in a formal manner on a special form so that it is easy to follow. An example of a borehole record sheet is given in Figure 5.3. The horizontal lines showing the divisions between strata are positioned to scale so that the thicknesses of strata can be seen at a glance. This is usually highlighted by a "legend" column which shows symbolically the soils encountered. The symbols used usually follow those recommended in BS 5930. The legend column is not simply a device to make the log look pretty but is useful for quickly comparing ground conditions in a number of boreholes. Typical scales used are 10mm or 15mm to 1m depth. This usually allows up to 10m depth to be shown on one sheet. For boreholes deeper than 10m, two or more sheets will be necessary. Trial pit records may be presented in a similar manner but are often given in the form of a scale sketch of a side of the pit. This latter method is more suitable where conditions vary from one end of the pit to the other. If necessary, two or more sides of the pit can be shown.

TEST RESULTS

Most laboratory test results are reported in tables. An exception to this is the grading results which are more usually presented as a grading curve, illustrated in Figure 4.3, since this is easier to assimilate than a row of figures.

Consolidation tests require the plotting of a number of graphs to obtain the m_v and c_v values. Graphs of specimen thickness against load, needed to obtain m_v values, are sometimes included. This enables the reader to calculate an appropriate m_v value for any pressure range. Graphs of specimen thickness against time, for each load increment, are necessary to obtain c_v values but these are not normally included in reports. Their inclusion would often make reports bulky and they would not be used by the majority of readers. However, interpretation of those graphs is not always easy and in certain circumstances the c_v values obtained may be questionable. It may therefore be desirable to include them sometimes.

The triaxial test also requires the plotting of

RECORD OF BOREHOLE Nº 26

Method of boring *Shell and auger* Diameter of boring *0.20m to 5.00m; 0.¹5m to 7.65m*

Ground level *31.20 m above O.D.* Details of casing *0.20m to 5.00m; 0.15m to 7.00m*

Samples and insitu tests Depth (m)	Type		Core recovery (%)	RQD (%)	Date and depth (m)	Level above datum (m)	Description of strata	Legend
0.50	*B*						*FILL (Grey-brown silty clay with stone and brick fragments)*	
0.60	*D*				*0.70*			
0.90	*S(17)*				*0.85*		*TOPSOIL*	
							Medium dense grey-brown slightly clayey fine to medium SAND	
2.20–2.65	*U*				*2.10*			
2.40	*W*						*Stiff dark grey sandy to very sandy silty CLAY with shell fragments*	
2.65	*D*							
3.20	*B*				*3.20*			
					3.50		*Grey-brown fine to coarse sandy GRAVEL*	
3.65–4.10	*U*						*Stiff dark grey very sandy silty CLAY with shell fragments*	
4.10	*D*							
4.90	*B*				*4.85*			
5.20–5.85	*U*				*5.10*		*Friable dark grey MUDSTONE*	
5.85	*D*						*Stiff to very stiff dark grey silty CLAY*	
					6.00 *28.8.81*			
6.70–7.15	*U*							
7.15	*D*							
8.20–8.85	*U*							
8.85	*D*				*8.85* *29.8.81*		*(End of borehole)*	

Key to symbols

U - 100mm dia. undisturbed sample
D - small disturbed sample
B - bulk disturbed sample
W - water sample
V - vane test
S(*) - standard penetration test
C(*) - cone penetration test
*N-value, blows/300mm

Remarks

Water was encountered at 3.20m depth and rose to 2.40m depth in ½h. It was sealed off by the casing at 3.50m. On the morning of 29.8.81 water stood at 2.15m depth.
The borehole was advanced by chiselling from 4.85m to 5.10m (2h chiselling).

Y. S. PARFITT LTD. | Job: Sedgethorpe development | Fig. No. 29

Figure 5.3 Example of a borehole log

graphs. Plots of axial stress against longitudinal strain must be drawn and Mohr circle plots must be constructed (plotting shear stress against direct stress) to obtain the shear strength parameters. Again, interpretation is not always easy and the Mohr circle plots should be included in the report, though too often they are not. Similar remarks apply to shear box tests. For the more specialist drained and consolidated undrained tests it is usual to also include the plots of axial stress against longitudinal strain.

RECOMMENDATIONS

Many site investigations are carried out for building projects where shallow strip or pad foundations will be used. Recommendations are given on allowable bearing pressure and foundation depth. These should be based on considerations of both ultimate bearing capacity and settlement and should be related to a specified foundation size. The ultimate bearing capacity of clays is independent of foundation size but larger heavier foundations will suffer higher settlements. For a given foundation load, the size of foundation may have little effect on the amount of settlement. This is because, although stresses are reduced, the depth of soil stressed is proportionately increased. For sands, the recommended bearing capacity must be associated with a given minimum size of foundation, since ultimate bearing capacity depends on the width of the foundation.

There is often confusion about what is meant by bearing capacity, since bearing pressures can be expressed in either net or gross terms:

(a) Gross bearing pressure is the actual pressure between the soil and the base of the foundation. This includes pressures due to the external loading of the foundation, the weight of the foundation itself and the weight of the soil above it.

(b) Net bearing pressure is the additional pressure between the soil and the base of the foundation, above that which originally existed. This includes pressures due to the external loading on the foundation and the weight of the foundation itself, but with an allowance for the reduction in pressure due to the weight of the soil displaced by the foundation. For spread foundations, which are almost invariably made of concrete with a density similar to that of the surrounding soil, it is usually sufficiently accurate to ignore the weights of foundation and displaced soil.

The distinction between net and gross bearing pressures is particularly important when considering deep hollow foundations, where the net bearing pressure will be considerably less than the gross value and may actually be negative.

The ultimate net bearing capacity (often called simply the ultimate bearing capacity) of a soil can be defined as the net bearing pressure imposed by the foundation when the ground beneath it is at the point of shear failure. The allowable bearing pressure is obtained by dividing this value by a suitable factor of safety (which should be specified in the report) then checking that settlement is acceptable. A further reduction may be needed if settlement is too high. No factor of safety is applied to settlement calculations. It should be made clear in the geotechnical report whether the allowable bearing pressure is limited by the risk of shear failure of the soil (i.e. the ultimate bearing capacity) or by settlement considerations. An estimate of settlement, and the time for settlement to be complete, should also be given.

The founding depth will usually be given as a minimum depth below the surface but it may be specified that foundations be taken to a particular level or be founded on a particular stratum. Factors which the geotechnical engineer should consider when recommending a founding depth include the following:

a. risk of shear failure (the ultimate bearing capacity can usually be increased by founding at greater depth);

b. suitability of strata (some layers may be weak or compressible);

TABLE 5.10
SUGGESTED PRECAUTIONS AGAINST CHEMICAL ATTACK OF BURIED CONCRETE

Class	Concentration of sulphates – as SO_3			Type of cement+	Min. cement content	Max. free water/cement ratio
	In soil		In groundwater			
	Total (%)	Water soluble* (g/1)	(g/1)		(kg/m³)	
1	<0.2	<1.0	<0.3	OPC, RHPC, PBFCC (plain concrete)	250	0.70
				(reinforced concrete)	300	0.60
2	0.2–0.5	1.0–1.9	0.3–1.2	OPC, RHPC or PBFC	330	0.50
				OPC, or RHPC with 70-90% slag or 25-40% pfa	310	0.55
				Sulphate resisting Portland cement (SRPC)	290	0.55
3	0.5–1.0	1.9–3.1	1.2–2.5	OPC, or RHPC with 70-90% slag or 25-40% pfa	380	0.45
				SRPC	330	0.50
4	1.0–2.0	3.1–5.6	2.5–5.0	SRPC	370	0.45
5	>2.0	>5.6	>5.0	SRPC + protective coating	370	0.45

*In 2:1 soil water extract. +OPC – Ordinary Portland cement; RHPC – Rapid Hardening Portland cement; PBFC – Portland Blast Furnace cement; SRPC – Sulphate-resisting Portland cement; pfa – Pulverised fuel ash.

Notes:

(1) This table applies to good quality concrete made with constituents complying with the relevant British Standard specifications and laid in near neutral ground conditions.

(2) Where OPC and RHPC are specified, they can be used with combinations of blast furnace slag or pulverised fuel ash.

(3) The minimum cement contents given include the content of slag or pfa and are based on 20mm nominal maximum size aggregate. For 10mm aggregate, increase the cement content by 50kg/m³ and for 40mm aggregate, decrease it by 40kg/m³.

(4) The percentages of slag and pfa are by weight of the slag/cement mixture.

(5) In severe conditions, such as thin sections or concrete subjected to high hydrostatic head, lower water/cement ratios or higher cement contents should be used. However, this may result in insufficient workability for thin sections, when a protective coating may be the only practicable solution.

(6) No special precautions are required for brickwork above the water table or in groundwater conditions with less than 0.3g/1 of sulphates. For groundwater sulphates of 0.3-3.0g/1, a 1:3 sulphate resisting cement:sand mortar should be used, with engineering quality bricks. Alternatively, a 1:¼:3 cement:lime:sand mortar may be used to improve workability.

c. consolidation settlement;

d. heave (foundations placed in or above the zone of seasonal moisture content fluctuations may suffer from expansion and contraction);

e. frost susceptibility (foundations should be placed below the zone of freezing and thawing in frost susceptible soils);

f. groundwater considerations (the presence of a water table at founding depth decreases the ultimate bearing capacity of sands and gravels by as much as a half; additionally, excavating below the water table for foundations may be difficult and costly);

g. cost (excavation and foundation costs increase with depth).

For piled foundations, suggested pile types, diameters and lengths are given, with details of allowable working loads. The allowable working loads of groups of piles may also be included, since the total capacity of a group of closely-spaced piles can be less than the sum of the capacities of the individual piles, particularly in clay soils. Information on the design parameters used should be given and supporting calculations should be included. Pile design is a specialist field and should be carried out by a geotechnical engineer. Most specialist piling contractors prefer to calculate pile load capacities themselves.

Buried structures may suffer from chemical attack due to impurities in the groundwater or aggressive chemicals forming part of the soil itself. The report should state whether chemical attack is a problem and, if necessary, should include recommendations on the precautions to be taken. In Britain these are usually based on recommendations given in BRE CP23/77, BRE Digest 250 and BS 8110:Part 1:1985. A summary of the main recommendations is given in Table 5.10.

Other recommendations in the report may relate to road pavement design, slope stability or other special problems. Where specialist topics such as these are dealt with, advice from the appropriate specialist engineer must be sought.

Chapter 6

GETTING THE MOST OUT OF THE GEOTECHNICAL INFORMATION

HOW THE GEOTECHNICAL ENGINEER REACHES HIS CONCLUSIONS AND RECOMMENDATIONS

A deeper understanding of the recommendations given in a report and their implications for the project can be obtained from an appreciation of the factors that the geotechnical engineer must take into account when organising and analysing the results. This is discussed in the following sections.

Decisions at the site work and laboratory testing stages

This has already been partly considered in Chapters 3 and 4 where the choice of site methods and laboratory tests was discussed in relation to the information required for design work. The decisions needed are reviewed below.

At the field work stage.

- What types of borings/trial pits/probes/in-situ tests should be used, how many are needed and where should they be located?
- What types of samples are needed, how many and from where should samples be taken?

At the laboratory testing stage.

- Which samples should be tested?
- What types of tests should be carried out?

Reviewing the information

As information reaches the engineer from the site and

laboratory he must look at it critically so that he can piece together a picture of the soil profile and the engineering properties of each soil type. This process can be considered in two steps:

- step 1: categorising the soils and drawing up an assumed profile based on borehole records etc.;
- step 2: grouping the soils together and deciding on design parameters for each soil.

Step 1. This primarily makes use of soil descriptions in borehole records, observations on site during site work and results of field tests (probes and SPTs). Laboratory test results (especially liquid and plastic limits) may also be useful to differentiate soil types. To draw up a profile, a chosen cross-section is drawn, showing the ground surface. The positions of all trial pits, boreholes and probes are marked on it, showing their depths, with the soil profile marked using abbreviated descriptions at each trial hole or probe. It may also be useful to include some test results such as SPT N-values and plasticity index values. For some tests, such as consolidation and shear strength tests, it may be more convenient simply to mark on the section where tests have been carried out, so the drawing does not become too cluttered.

A single section may be sufficient for linear problems such as a road alignment or a slope stability analysis but, where the investigation covers an extended area, several sections may be needed. A useful dodge, to overcome the problem of envisaging the soil profile in three dimensions, is to plot the sections using an isometric projection, as illustrated in Figure 6.1.

Step 2. Once the soils have been categorised into groups then the test results for each soil type can be looked at with a view to deciding on suitable values for use in design. As with the soil profile, it may be useful to plot out results. For example, shear strength results from several boreholes may be plotted against depth for a particular soil type, as illustrated in Figure 6.2. A line indicating design shear strength against depth can then be drawn on the diagram. Where large numbers of results are available, those so inclined can carry out a regression analysis

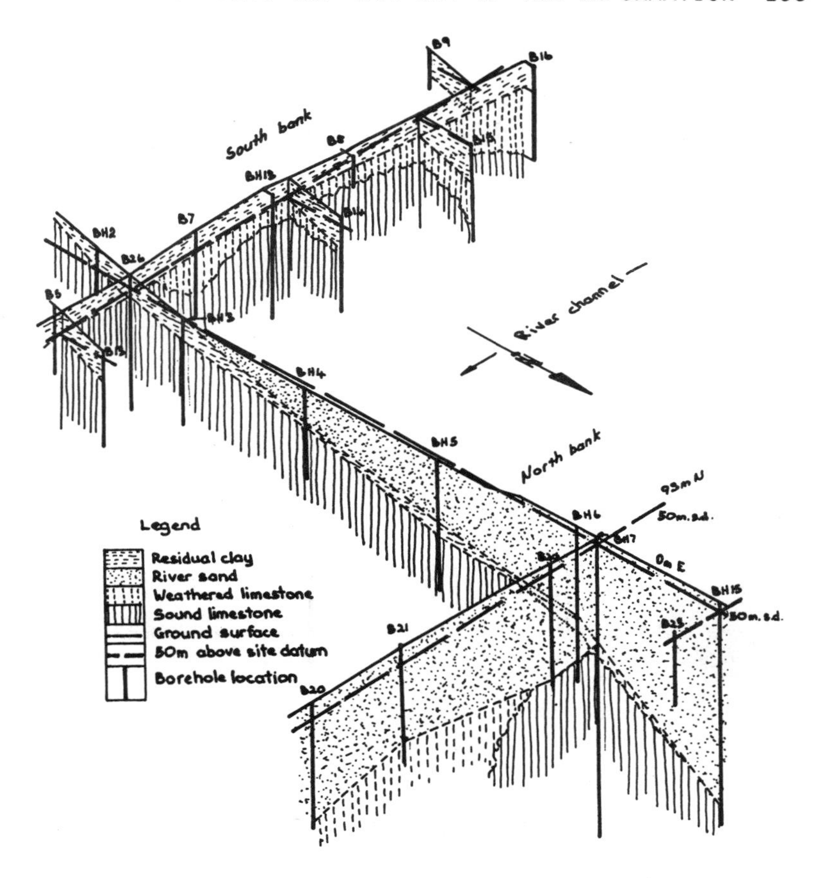

Figure 6.1 The use of isometric projection to depict ground conditions in three dimensions

and plot the design line at a given number of standard deviations (usually two) below the average. However, it is adequate for most purposes to eye-in a line which falls below most of the test results, as illustrated. The design line must be sufficiently conservative to ensure the design is safe, even allowing for the odd patch of weak ground, but not so conservative that the design becomes unnecessarily costly. When results are plotted out in this way, variations in values over different parts of the site may become apparent and the engineer must decide

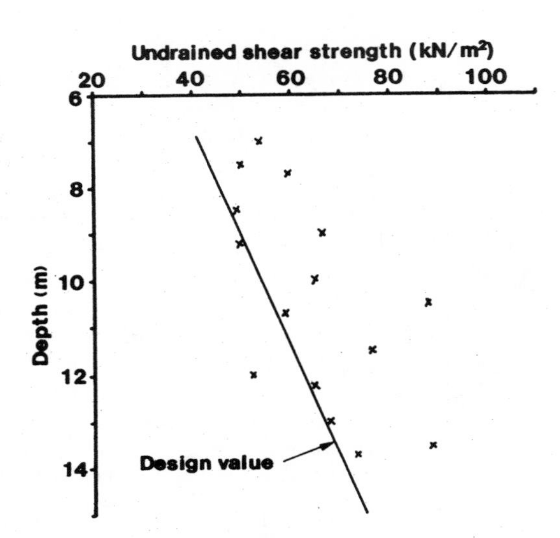

Figure 6.2 Example of a plot of shear strength against depth, used as an aid to selecting design values

whether one design value can be used for the whole site or whether different values are appropriate in different areas. Also, the plot will show up exceptionally high or exceptionally low values. For these, the sampling and testing records should be checked to see whether there is any explanation for the abnormality. Exceptionally low values are particularly worrying because the engineer must decide whether this is a freak result, perhaps caused by the sampling or testing method, or whether it indicates areas of very weak ground: his decision will affect the design line chosen.

SPT N-values may also be plotted against depth and treated in a similar fashion. However, they tend to be more erratic and a rule-of-thumb often used is simply to calculate the average value for a given stratum in each borehole and use the minimum of these averages as the design value. This simple approach is not suitable where N-values vary significantly with depth. Corrections are usually applied to values where tests are carried out in saturated silts. Some engineers also apply depth corrections to results at shallow depth but these are somewhat contentious and should be applied with care or not used at all.

Consolidation tests present a more complex problem. Not only will the coefficient of volume compressibility, m_v, (which gives the amount of settlement) vary from sample to sample but within each sample it will vary with consolidation pressure. Also, from the design point of view, we are concerned not simply with the maximum probable value of m_v, to give the maximum likely settlement, but also with the probable range of m_v values beneath a structure. This is because variations in compressibility can result in differential settlement which is usually of more concern than total settlement because of the resulting distortions to structures. The usual method of treatment is to compare m_v values from several samples of the same soil type, at equal consolidation pressure ranges. The consolidation pressure range chosen is equal to the expected pressure range at the site. Thus, for a typical foundation problem, the pressure range will be from overburden pressure (at the depth being considered) to overburden plus the excess pressure caused by the foundation loading. Settlement calculations are too imprecise to enable values to be confidently predicted to a high degree of accuracy. Additionally, for many sites, it is impractical in terms of cost and available time to carry out a sufficient number of consolidation tests to allow suitable design values of m_v to be accurately assessed.

The rate of consolidation may also be important. For instance, if most of the settlement of a building takes place before surface finishes are added then settlement cracks are less likely to cause problems than if settlement were to occur over several years. The required time for substantial completion of settlement can be calculated from the coefficient of compressibility, c_v. Like m_v, its value decreases with pressure but the variation is less marked. A brief inspection of the c_v values for the appropriate pressure range is usually all that is needed to arrive at a suitable design value. A more detailed appraisal is not justified for two reasons. Firstly, c_v values obtained from the consolidation test are often not representative of the true values for the soil mass, as discussed in Chapter 4. Secondly, the overriding factor in the rate of consolidation is the length of the drainage path (the distance water must travel to the ground surface or to more permeable material when squeezed out of the soil) and, often, this can only be guessed at.

A good geotechnical report should not simply state the design values used but should include the geotechnical engineer's reasons for choosing these values, with reference to appropriate tables or graphs.

Using the information for analysis and recommendations

Having decided on design values for the soil properties, the next step is to translate these values into recommendations for the designers. This requires an analysis of the geotechnical problems, which cannot be carried out until details of the project have been decided. For instance, what types of foundations are to be used and what loads must the foundations carry? Will retaining walls be needed and to what height? What heights of embankments are needed? Are there any special constraints on the surrounding ground movements that can be tolerated? On the more straightforward projects the answers may be obvious; on others there may be a need for discussions between the geotechnical engineer and others on the design team. It may be necessary to analyse a number of possible options.

For foundation design, the first consideration is the depth to the bearing stratum. Where a good bearing capacity can be obtained at a shallow depth, strip or pad footings will usually be chosen. On grounds of cost, these should be kept as shallow as possible but they may need to be made deeper for a variety of reasons, such as those given below.

(a) The upper layer of soil may be of poor bearing capacity; too weak or compressible. All foundations must be taken below ground surface to some extent so that they are not bearing on topsoil, and to avoid the risk of soil being excavated below the founding level alongside the foundation.

(b) Increasing the foundation depth increases the

bearing capacity of spread foundations. This method of increasing bearing capacity should not be relied on too heavily for shallow foundations because there is always the risk that somebody will dig a trench or regrade the ground alongside a foundation, removing the advantage of depth.

(c) The upper layer of soil suffers seasonal variations in moisture content. Some soils shrink when they dry out and swell when they become wet again. An example is London Clay. With these soils, foundations should be taken below the depth of significant seasonal moisture content changes. This varies with soil type but, in Britain, is typically 1.5-2m. This is discussed in Chapter 7.

(d) Some soils show pronounced swelling when frozen - frost heave - and then become very weak on subsequent thawing. With these soils, foundations must be taken below the expected level of frost penetration. This is discussed more fully later and in Chapter 7.

Once the founding depth has been decided, the ultimate bearing capacity can be calculated: methods for this are given in Chapter 11.

In clay soils the allowable bearing capacity is initially assumed to be equal to the ultimate bearing capacity divided by a suitable factor of safety (typically 2½ or 3), as discussed in Chapter 1. Settlement calculations should then be carried out. If the calculated settlements are too high then the value of allowable bearing pressure must be reduced to keep them within acceptable limits. It is not usual to use a factor of safety for allowable bearing pressures based on settlement limits.

Where ultimate bearing capacity is the overriding consideration the problem is normally fairly straightforward: the foundation is made big enough to spread the load sufficiently to reduce the bearing pressure to the required value. In some cases it may be necessary to consider alternative foundation types. Where settlement is the main consideration, simply increasing the foundation size to reduce the bearing capacity may be of little help: this is discussed in Chapter 11. There are several possible solutions.

Where the compressible layer is not too thick it may be possible to transfer loads through it by simply using deeper foundations or by resorting to piles. The problems caused by a building founded on a thick layer of compressible soil are particularly difficult to solve and should never be underestimated. Raft foundations may be used to reduce differential settlements; or hollow basements may be used to reduce loading, hence total settlements, and to provide a rigid base. Such solutions are expensive and may not be considered justifiable, especially for small structures. However, if it is decided to found on soft, compressible material then there will be a penalty to pay in terms of high total and differential settlements, and design details must allow for this. This is discussed in Chapter 7. Whichever solution is chosen it will usually have repercussions for the design of the structure as well as the foundations: the structural designer or architect must become involved and cannot simply assume that settlement problems can be left entirely to the geotechnical engineer.

In granular soils, foundations must similarly be designed to an allowable bearing pressure which has been obtained from considerations of both ultimate bearing capacity and settlement. Because the ultimate bearing capacity of granular soils is usually very high, restriction of settlement is almost always the overriding factor in the design of foundations on such soils: so much so that ultimate bearing capacity is sometimes not even checked. This is a dangerous practice because there is always the chance that the ultimate bearing capacity may be exceeded, especially with narrow foundations on loose sands. Settlements in granular soils are usually rapid because of their high permeability. It is normally assumed that settlement will take place as construction proceeds so that it will normally be complete before surface finishes are added. Because of this, and the difficulty of sampling and testing, specific tests to determine the rate of consolidation are not normally carried out.

As foundations settle, the soil beneath them is compressed and gains in strength. Thus, they are least stable just after they have been built, before there has been time for water to be squeezed out of the soil and settlement to take place. This immediate response of the soil, often referred to as the "end of construction condition", is simulated in the quick undrained triaxial test. The test is quick and easy to perform, and therefore cheap. Also, any changes in groundwater pressures due to the added loading are automatically reproduced in the triaxial test, so there is no need to worry about what happens to the pore water pressures: only the overall response of the soil is considered, making the analysis relatively simple. This is discussed in Chapter 10.

With excavations, the situation is the reverse of that for foundations because the removal of soil causes a reduction in pressure. In the long term, the soil swells and loses strength. Thus, the sides of the excavation become less stable with time and it is the long-term conditions which must be considered for stability analysis. The stability depends critically on the long-term groundwater pressures which themselves depend on the drainage conditions in the surrounding soil and these cannot be simulated in the triaxial test. Because of this, the responses of the soil skeleton and the soil water to applied loads must be considered separately, both at the testing stage and during analysis. This is discussed more fully in Chapter 10 but it can be seen that problems involving the stability of excavations are usually complex and require specialist knowledge.

The stability of natural slopes and embankments is analysed in a similar manner, considering long-term conditions and taking into account the pore water pressures in the soil. However, the stability of large embankments can be yet further complicated by high pore water pressures which can be built up during construction as the soil is compacted, resulting in short-term instability. If a stability analysis shows this to be a problem then it may be necessary to monitor pore water pressures during construction and to limit the rate at which construction proceeds.

Water pressures behind earth retaining walls are also important when considering their stability. Where walls retain a granular material there is no difference between short-term and long-term stability because the high permeability allows long-term water pressures to be rapidly established. However, if clay backfill is used, the low permeability prevents rapid establishment of equilibrium pore water pressures and both the short-term and long-term stability must be considered because either could prove to be more critical. The stability of earth retaining walls and slopes is discussed more fully in Chapter 12.

The design of road pavements is again a specialist field. In Britain, the structural design of the pavement is carried out according to Road Note 29, issued by the Transport and Road Research Laboratory. Practices vary in other countries but, where English is the technical language, the American AASHTO method is commonly used as a basis for design procedures. In Commonwealth countries the TRRL Road Note 31 is also used for minor roads. The TRRL methods use soil strength values based on the California Bearing Ratio test, described in Chapter 4. The AASHTO method does not specify what soil strength test is to be used but, in practice, the CBR test is usually chosen. In Britain it may not be necessary to carry out CBR testing even for projects that include road construction: Road Note 29 contains a table giving correlations between CBR value and plasticity index, so the much simpler and cheaper plasticity tests can be performed instead.

PROBLEMS ARISING FROM LACK OF KNOWLEDGE

As discussed in Chapter 1, a great many of the problems that arise out of deficiencies in the geotechnical report are not the result of errors or lack of judgement. They simply reflect the fact that even after a thorough site investigation has been carried out the knowledge of ground conditions is far from complete. The soil profile is known only at the borehole and trial pit positions: in between, we have to rely on guesswork. It can be instructive to note the variation in ground conditions that can occur across a single trial pit and to try to imagine what interpretation would have been put on the results if a borehole had been sunk instead. Would the assumed soil profile have been the same? When using the recommendations given in a geotechnical report, the accompanying warnings about unforeseen conditions should always be heeded and passed on to site staff during construction.

The report recommendations may also be limited by the lack of knowledge of a different kind. The geotechnical engineer may not have been given sufficiently detailed information on the proposed development to enable him to give more than general advice. At worst, the recommendations may even contain some implicit assumptions that are not valid. Often, of course, decisions regarding the types of construction or even the overall layout will not have been made when the site investigation is carried out. Nevertheless, the geotechnical engineer will always be in a better position to tailor his recommendations to the requirements of the designers if he is given the maximum possible information on the proposed development.

LOOKING FOR WARNING SIGNS

It is by no means unknown to find that a site investigation has revealed evidence of ground conditions which could cause problems, but that it has been ignored in the design. This is not out of a conscious decision to ignore the potential problems but because they simply have not been noticed; or, if they have, their implications have not been realised. All too often this results in difficulties during construction or long-term problems with structures, such as cracking of walls or movement and distortion of floors.

Perhaps the difficulty in picking up warning signs arises because it is the interaction between the structure and the underlying soil which causes the problems. As a result, the geotechnical engineer responsible for the report may not properly appreciate the consequences of the ground conditions for the proposed structures, and the structural designer may similarly not appreciate the significance of the test results given in the geotechnical report. Even where the geotechnical engineer does comment on potential problems, his remarks may not be fully appreciated by the designer, whose interests and expertise lie elsewhere. To avoid this problem the structural designer should look a little more closely at the report than simply reading the recommendations. Without becoming a soils expert himself, and without spending too much time on the report, he can check for warning signs in the test results and, if necessary, refer back to the geotechnical engineer. How to interpret test results, to get a feel for the their values, and to pick up pointers to potential problems are discussed in the following sections.

Shear strength

Clay soils are categorised according to their shear strength as indicated in the first column of Table 5.2. Soils in the soft or very soft categories will obviously give stability problems for all but the very lightest structures. In addition, low strength soils may have other undesirable properties, such as high compressibility.

Compressibility

The compressibility of clays is measured by m_v, the coefficient of volume compressibility. Table 5.5 shows descriptive terms that are used and indicates typical compressibility values for various soil types. Settlement problems are likely to arise in soils with high and very high compressibility, and may occasionally arise with soils of medium compressibility. Highly compressible soils may also exhibit undesirable shrinkage/swelling characteristics with seasonal changes in moisture content, although this problem is not limited entirely to soils of high compressibility, as discussed below and in Chapter 7.

Plasticity

The liquid and plastic limits and the plasticity index, described in Chapter 4, give valuable indications of the type of behaviour that can be expected from a clay and, consequently, are used to classify clay soils. The significance of the test results is most easily appreciated by use of the Casagrande plasticity chart shown in Figure 5.2, which is a plot of plasticity index against liquid limit. Results tend to plot a little above or below the A-line, as indicated by Figure 6.3.

Soils with a liquid limit greater than about 50 are likely to be weak and compressible, although these properties are better indicated directly by strength and compressibility measurements. What makes a high liquid limit particularly worrying is that it may indicate an expansive soil: one which expands as moisture content increases and shrinks as the soil dries out. This movement is caused entirely by changes in moisture content and is not related to consolidation settlement. The presence of an expansive clay may also be detected by a plasticity index greater than about 20 as indicated by Table 6.1. The problem may not develop for many years; for instance, until a particularly dry summer reduces the moisture content at depth causing shrinkage of the soil and cracking of structures, or until a burst pipe under a building causes the soil to swell.

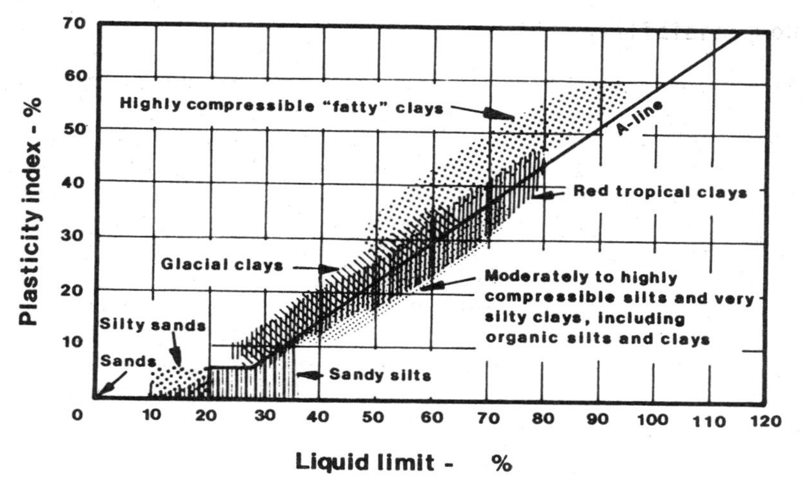

Figure 6.3 Typical ranges of liquid limit and plasticity index for various soils.

If an expansive clay is suspected, an obvious check is to examine buildings in the surrounding area to obtain as much information as possible on the foundation types and depths used and to note any problems of cracking. A number of tests have been

TABLE 6.1
APPROXIMATE RELATIONSHIP BETWEEN PLASTICITY INDEX AND INHERENT SWELLING CAPACITY

Plasticity Index (%)	Inherent swelling capacity
0-15	Low
10-35	Medium
20-55	High
35+	Very high

devised to measure the swelling potential of clays but it is not possible to predict what movements will occur: the best that can be hoped for is to identify the problem and design to avoid it. A useful test which can be used as an added check on a suspect soil is the linear shrinkage test. A specimen of soil is mixed with water so that its moisture content is equal to the liquid limit and the resulting paste is spread into a small trough. This is oven dried, causing the soil to shrink. The linear shrinkage is defined as the reduction in length of the specimen expressed as a percentage of its original length. A linear shrinkage greater than 8 indicates a clay with a high swelling potential. Marginal problems could be experienced with a linear shrinkage as low as 5.

Moisture content

You do not need to be a geotechnical engineer to know that drier soils are generally stronger and stiffer than wetter ones. However, to be a little more precise than this, it is useful to relate the natural moisture content of a soil to its liquid and plastic limits. This can be done by calculating the "liquidity index" which is defined as,

Liquidity index = (m - PL)/PI

where m is the natural moisture content, PL is the plastic limit and PI is the plasticity index.

Thus, the liquidity index will be 0 for a soil at the plastic limit and 1 for a soil at the liquid limit. A soil with a liquidity index of 0.5 will have a moisture content which is mid-way bewteen the liquid and plastic limits. The liquid and plastic limits are, in fact, perverse forms of shear strength tests: all soils at the liquid limit have a shear strength of about 1kN/m^2 and all soils at the plastic limit have a shear strength of about 100kN/m^2. Thus, the liquidity index gives an indication of the strength of the soil: a value approaching 1 is a warning that the soil is almost in a liquid state.

Chemical attack

The presence of aggressive chemicals from industrial waste and like materials is something which must be taken into account, but it is a specialist topic beyond the scope of this book. However, the most common sources of chemical attack on foundations are organic substances and natural sulphates in the soil itself. Specific tests should be carried out if chemical attack is considered to be a potential problem and the appropriate measures taken, as described in Chapters 4 and 5. Organic soils are a particular problem and it is important that they be identified as such. It is impractical and unnecessary to test all soils for organic content but organic soils can be detected by other means: considering a soil's origin; visual inspection; and classification using the Casagrande plasticity chart, Figure 5.2. For instance, estuarine silts and clays are likely to be organic, and organic soils are usually dark in colour and may have a distinctive smell. They usually have high liquid and plastic limits and plot below the A-line on the Casagrande chart, as indicated in Figure 6.3.

Frost susceptibility

Frost susceptibility is the tendency of a soil to heave and weaken during cycles of freezing and thawing. In clays, the water in the adsorption complex (surrounding the clay particles) freezes at a lower temperature than the free water in the pores so that both liquid water and ice can exist together. The freezing out of water from the soil tends to draw in more water through the still-liquid adsorbed layer. This increases the soil moisture, allowing more ice to form, with consequent further heaving and a dramatic loss of strength on thawing. The phenomenon is less marked in heavy clays because their very low permeabilities do not permit the ready flow of water. At the other extreme, granular soils, though highly permeable, do not possess an adsorbed layer which remains unfrozen and are usually well drained, so frost heave and loss of strength are less of a problem, although problems can occur if there is a source of water under pressure. Frost susceptibility then, tends to be a feature of the more permeable clays; that is, those of low and medium plasticity. Table 6.2 gives a correlation of frost susceptibility with grading and plasticity index, suitable for the preliminary identification of soils, based on recommendations by the Transport and Road Research Laboratory.

TABLE 6.2
FROST SUSCEPTIBILITY OF SOILS

Permeability rating	Identification	Frost Susceptibility
High permeability	Granular: <10% finer than 75μm	Not susceptible
Intermediate permeability	Granular: ≥10% finer than 75μm	Susceptible
	Cohesive: PI< 20	
Low permeability	Cohesive: PI≥ 20	Not susceptible

Notes: In addition to soils identified from the table above, the following materials are known to be frost susceptible:

- all chalks.
- burned colliery shales.
- oolitic and magnesian limestones with a saturation moisture content greater than 3 per cent.

With rocks and coarse aggregates, frost susceptibility increases with saturation moisture content.

Variations in ground conditions

If soil conditions change from one part of a structure to another, differential settlement may be a problem. Where changes are abrupt, considerable distortions may occur in parts of the structure. This can happen because of sloping soil strata, sloping ground, variations in thickness of soil strata or, perhaps most dramatically, where a structure is partly on natural ground and partly on fill.

Filled ground

As well as the problem of a structure on the border of a filled area, the fill itself can have undesirable properties. It may be very variable, poorly compacted, or may settle with time or with changes in the height of the water table. Properly laid and compacted fill that has had time to settle may provide a sound foundation but, in general, building on filled ground is a hazardous business which requires extra precautions. An added problem, which arises with fills composed only of locally-derived natural soils, is that of identifying what is fill and what is undisturbed natural soil. Some of the special problems of fills are discussed in Chapter 7.

Other hazards

Certain areas have their own hazards which are not due to the properties of the underlying soils or rocks but are nevertheless important. These include mining subsidence, flooding and coastal erosion. It is surprisingly easy to overlook these more global problems while coming to grips with the details of the soils investigation. The problems of mining subsidence are considered in Chapter 8.

USE OF THE REPORT AT THE DETAIL DESIGN STAGE

As discussed previously, where problems arise because of poor foundation conditions, many of the solutions lie in the hands of the structural designer rather than the geotechnical engineer, especially if solutions are to be as cost effective as possible. Even where special foundations are provided, attention to detailing may still be important. For instance, deepening the foundations may help reduce settlement of a building but if the ground floor slab is laid directly on a soft, compressible clay then considerable movement of the floor slab may result. It may be that the expected settlement will not pose a structural problem but will result in unsightly cracking of finishes. In this case the best solution may be simply to apply more flexible finishes. Simple devices such as stiffening strip foundations with reinforcing rods where they cross from one soil type to another, or providing specially stiffened extra shallow foundations may bring significant improvements for little extra cost. The solutions will vary from site to site but, where soil conditions are poor, significant improvements can often be achieved by careful attention to design details. This is discussed in more detail in Chapter 7.

Chapter 7

PROBLEM SOILS

It can be said that all soils present their own particular problems so far as construction is concerned and that few construction projects of any size encounter no difficulties with ground conditions. But most of the problems encountered are relatively straightforward and are overcome using established practice and commonsense. On some sites, however, the nature of the underlying soils will give rise to particular difficulties that require special treatment and exceptional care.

This chapter reviews the types of problem soils that may be encountered, describes the nature of the problems and discusses their consequences for construction. Many of the problems described here are also covered in other parts of the book, as the numerous cross-references indicate. However, such a disproportionate number of foundation failures occur in areas underlain by problem soils that a special chapter devoted to this topic seemed appropriate, even though it inevitably leads to a certain amount of repetition.

HIGHLY COMPRESSIBLE SOILS

The problems of building on highly compressible soils have already been introduced in Chapter 6, in the context of the site investigation report. Soils in this category are soft clays, organic clays and silts, and peats. They tend to occur in flat, low-lying areas such as the fens of Lincolnshire or the Somerset Levels, estuarine areas and old lake beds, although peat and soft organic clays are also found in high, poorly-drained moorland. Soils are usually easily identified and the problem can be defined quantitatively in terms of the coefficient of volume compressibility m_v, measured in the oedometer test, described in Chapter 4. Settlement estimates can be obtained as described in Chapter 11.

Treatment for relatively thin layers of compressible soil may include taking spread foundations through the layer, or providing piled foundations. If piles are used, it may be necessary to allow in the design for negative friction; the downdrag on the upper part of the pile caused by consolidation of the compressible layer. This is discussed in Chapter 11. Attention to structural detailing is important. For instance, floor slabs laid directly on a compressible layer are likely to distort and settle relative to the better-founded walls. Interior partition walls with shallow foundations will suffer similar problems. Framed buildings that have external cladding or infill panels of blockwork or brickwork on shallow foundations can look particularly unsightly as the frame and cladding go their own separate ways. Window and door openings add to the problem by creating variations in contact pressure beneath shallow wall foundations, causing additional distortions and cracking within the walls themselves.

Not only are buildings affected, but also access roads and hardstandings. Differential settlements of a metre or more can occur, with dire consequences to both the serviceability and the structural soundness of the pavement. In some cases, embankments and areas of fill can settle several metres.

The problem is particularly intractable where soft deposits are of great thickness. High settlements have to be accepted, as discussed in Chapter 6 and Chapter 11, and it is especially important to consider the effects on the whole structure. Some special measures which may be used are indicated below.

Surcharging or special ground treatment

Consolidation can be speeded up by placing surcharge loading in the form of extra fill, over an area. This is then removed just prior to construction. Its effectiveness depends on the thickness of material placed, the length of time it is left in place, and the thickness and permeability of the underlying compressible layer. Several metres of fill may need to be left over an area for several months.

Vertical drainage can be used beneath fill or heavy foundations, or in conjunction with surcharging, to speed up the rate of consolidation. A special machine pushes the drain - a plastic former wrapped in filter fabric - into the ground. Alternatively, vertical sand drains may be used.

Other methods of treatment include dynamic consolidation, in which a large weight is dropped repeatedly to create very high impact forces; and, for sands, vibro-compaction (or vibro-flotation), in which a large vibrating "poker" is inserted into the ground. The vibro-compaction method may also be used in conjunction with the "stone-column" technique, in which stones are poured into the hole created by the vibrating poker. A treatment used for clays in Scandinavia and a number of other countries is known as the "lime column" method, or "deep stabilization". A special auger bores through the clay and, once it has reached the required depth, is slowly withdrawn whilst injecting quicklime into the clay and mixing it up, to produce a column of stronger, stiffer, less plastic clay. In general, the types of treatment described above require specialist knowledge and are practicable only for major projects.

Use of raft foundations

The use of reinforced concrete rafts of the types shown in Figure 2.4 will considerably reduce distortions in structures, though at increased cost. High settlements and overall tilting of the foundation and structure may still occur, possibly creating problems with services, especially drains and sewers. Settlement of the building relative to the surrounding ground may also cause difficulties.

Use of buoyancy rafts or basements

If the weight of soil removed is about the same as the weight of the structure and its foundation, there will be little or no change in overall pressures within the soil so that settlement will be much reduced or, in theory, eliminated altogether. The cost of providing such foundations is high but against this must be balanced the extra amenity value of the space created, the rigidity which the foundation provides, and the fact that this may be the only method of reducing settlements to tolerable values where deep deposits of soft clay are encountered. Speed of construction may be critical: once the basement has been excavated, the ground beneath will tend to swell in response to the reduction in loading; if a sufficient time elapses for a significant amount of swelling to take place then some of the advantage will be lost.

Use of lightweight construction

This is a common practice on industrial estates which are often built over poor ground, even on old refuse tips. Steel construction with lightweight flexible cladding reduces settlements and is more tolerant of differential settlement than more traditional structures. For those who find it acceptable, timber frame construction may be used for housing, to keep down weight, although the use of traditional plaster finishes and brick or render outer walls reduces some of the advantages.

Delayed application of finishes

In many cases, the distortions caused by differential settlement are not a threat to the stability of the structure: the main problem is the damage to surface finishes such as plastering and rendering. If it is estimated that settlement will be substantially complete after several months rather than years, then a delay of a few months before completing surface finishes can reduce subsequent damage. The advantage gained is likely to be better than anticipated because most settlement takes place quicker than the estimated time, for the reasons discussed in Chapter 4 and Chapter 11.

Detailing of floors and non-loadbearing walls

Problems of differential movement between elements of a building can occur where floor slabs and the shallow foundations of non-loadbearing walls rest on a compressible soil, whilst the foundations for the main structure are taken down to a good bearing stratum. Solutions are numerous. They include use of lightweight partitioning instead of blockwork for internal walls, and external cladding that is supported by the main framework of the building. Ground floors may be suspended but this is often considered too costly. In the long term, however, it may be the best solution taking account of the inconvenience and maintenance costs which result when differential movement takes place. An alternative is to excavate the clay beneath the floor area and replace it with compacted granular fill, although this may prove less satisfactory where compressible soils extend to some depth.

EXPANSIVE SOILS

These are soils that swell or shrink in response to changes in moisture content. In such soils, building movements may arise not only from consolidation settlement due to the applied loading, but also from changes in the soil moisture brought about by factors such as seasonal weather variations or changes to the pattern of groundwater flow. Expansive soils are intially identified by their liquid limit and plasticity index as discussed in Chapter 6 and indicated in Table 6.1. The linear shrinkage test, also described in Chapter 6, forms a useful check, and there are a number of other specialist tests which quantitatively measure the swelling potential and swelling pressures induced in such clays. However, it is not possible to predict what movements will occur, even with the most sophisticated tests.

In Britain, expansive properties tend to be associated with soils of high compressibility. Problems are caused by shrinkage rather than expansion. Buildings may show little or no distress until an exceptionally dry summer dries out the clay, which shrinks. The extent of drying, hence shrinkage, usually varies from one part of the site to another, causing differential settlement, hence cracking. The problems of differential drying may be made worse by the presence of trees which help dry out the soil in their vicinity. Fast-growing trees, such as willows and cypresses which consume a lot of water, are a particular problem and should not be planted near buildings.

Opposite problems, caused by wetting of the soil, can result from occurrences such as a burst water pipe or sewer, the channelling of rainwater into the foundations, or the removal of a long-established tree close to the building. Wetting of the soil does not usually lead to significant expansion in Britain, where soils are mostly already saturated, but severe difficulties can arise in warmer, drier regions. For those involved in the construction of buildings on expansive soils in arid or semi-arid regions, swelling/shrinkage considerations can be a major factor in design. In these conditions, a burst pipe, watering of gardens or poor attention to rainwater drainage can have disastrous effects on buildings.

Remedial measures in Britain are usually limited to taking foundations below the zone of significant seasonal moisture content variation (typically around 1.5m depth) and using suspended ground floors or replacing the upper layer of soil beneath floor areas by compacted granular fill. Much more elaborate precautions may be needed in arid or semi-arid regions. Some of the precautions used to prevent damage are outlined below.

Spread foundations

These are taken below the zone of seasonal moisture content variation, provided it does not extend to excessive depth. Expanded polystyrene or loose sand is placed at the sides of strip foundations, especially on the inner faces, to reduce any lateral thrust resulting from expansion of the surrounding clay. Expansive soils in dry regions are often strong, with low consolidation characteristics, and appear superficially to make an excellent foundation. Full use is made of these properties by keeping footings as narrow as possible to increase contact pressures: this is said to help in a minor way to prevent the clay beneath the foundation from swelling.

Short bored piles

These are often used in place of spread footings, which become less economical as depth increases. The pile is designed to resist uplift which results from swelling of the soil in the zone of seasonal moisture content variation. Alternatively, special precautions may be taken to reduce uplift forces, such as surrounding the upper part of the pile in polythene, as illustrated in Figure 2.5. The pile capping beam is cast clear of the ground surface.

Ground floors

Suspended floors, or the replacement of the upper part of the clay by selected backfill, as described previously, are used. Groundbearing floor slabs are sometimes built with a honeycombed underside, produced by placing formers, such as split-in-two plastic pipes, on the ground before pouring the concrete floor. This limits contact between the slab and floor to a small proportion of the total floor area and provides an underfloor space for the clay to swell into. The success of such systems is doubtful. With any form of groundbearing slab, slip joints must be provided to allow movement between the slab and the walls.

Drainage

Special care is taken to lead rainwater well away from the foundations. This is in contrast to normal practice in many tropical and subtropical areas where rainwater from the roof is allowed to cascade freely to the ground. Water pipes and sewers may be laid in drainage channels so that, should a leak occur, water will drain away from the building rather than seep into the soil.

Moisture barriers

Buildings may be surrounded by vertical moisture barriers, typically polythene sheets, which extend to below the zone of seasonal moisture content variations. Alternatively, aprons may be used to reduce moisture content changes beneath buildings. These typically extend about 1.5m from the walls and are made of concrete, asphalt or buried polythene sheeting. In addition to seasonal variations in moisture content there is a gradual migration of moisture towards buildings caused by the presence of the building itself, rather like the dampness beneath a stone. Although impermeable barriers prevent seasonal moisture variations beneath buildings, they merely delay this long-term increase. To counteract it,some engineers suggest the use of semi-permeable aprons, made of lean-mix concrete or open-textured macadam. These channel surface water away from the building but allow evaporation through their surfaces.

Replacement of soil

The soil beneath a proposed building is removed to the depth affected by moisture content changes and the excavation filled with compacted granular backfill. The depth of soil replaced should be typically 1m to 1.5m below the underside of the floor slab or foundations. The soil replaced should extend at least 2m beyond the edges of the building and it is imperative that the bottom of the open excavation is not allowed to become excessively wetted. The cost of this may be offset against savings from the simplified floor slab and foundation requirements. As an alternative to using granular backfill, the soil itself may be used, stabilised by mixing it with 3 to 8 percent lime. However, experience with this technique is limited.

Deep basements

These take the foundation depth below the worst affected surface layer and, provided a rigid reinforced concrete box construction is used, give added resistance to warping and cracking. Basements should be surrounded by granular backfill to help protect them from lateral thrusts.

Pre-wetting

The logic behind this approach is that, since the moisture content beneath buildings tends to rise and maintain a higher equilibrium value in the long term, then it is better to produce this rise, with the attendant swelling, before construction starts. This is achieved by constructing a low embankment round the area and flooding it. Ponding of the area for at least a month is practised in some regions.

COLLAPSIBLE SOILS

These are soils that have an open structure, with a high voids content. Such soils are typically free-draining and have been rarely, if ever, saturated. Wetting the soil under loading weakens the bonds between grains and allows them to settle into a more compact arrangement. Examples of naturally-occurring collapsible soils are loess and some red coffee soils. The phenomenon is not usually a problem in temperate climates but it can occur in certain types of fill, particularly open-pit mineworkings backfilled with loosely-compacted overburden or with mining waste products.

Initial identification of collapsing soils relies on the experience of the geotechnical engineer, and it is easy to overlook the problem. The extent to which a soil will collapse on wetting can be judged by running a special series of consolidation tests. A number of identical samples are tested but, instead of the usual procedure of flooding the specimens at the first load application, only one specimen is flooded and the other samples are left dry. Each is flooded at a different load increment so that the effect of flooding the soil at various confining pressures can be ascertained.

Remedial measures are very difficult. Many of the measures described for expansive soils are applicable. The use of dynamic compaction or vibro-compaction, described for compressible soils, may be appropriate. One Russian solution to the problems of loess is to induce compaction by the simultaneous detonation of charges at the bottom of a pattern of boreholes!

SENSITIVE CLAYS

Generally, if a piece of clay is moulded in the hand it

will become softened: its shear strength will decrease. This decrease is known as sensitivity and is expressed quantitatively by the coefficient of sensitivity, S, defined as

$$S = \frac{\text{natural shear strength}}{\text{remoulded shear strength}}$$

With most British clays, the loss in shear strength is relatively unimportant. Even after substantial remoulding, values of S betweeen 2 and 4 are common. However, clays in some areas, notably parts of Russia, Scandinavia and Canada, are highly sensitive, with values of S exceeding 100. To cope with this, local construction practice is adjusted to minimise soil disturbance.

The phenomenon is complex but appears to be associated with soils that were laid down in saline conditions. Like collapsing soils, they have an open structure. This formed, and was maintained, because of electrical charges on the surface of the clay particles as they settled out of suspension. Subsequent gradual leaching of the cations within the clay minerals, as the soil moisture changed from saline to fresh, altered the electrochemistry of the particles and left the soil in an unstable condition: the open structure being maintained virtually out of habit. Any disturbance can break the flimsy bonds and cause a loss in strength. Again, like collapsing soils, the particles may try to settle into a denser packing. However, soils in these northern regions are usually fully saturated so the particles are prevented from packing closer together by the water filling the pores which, because of the soil's low permeability, cannot drain away rapidly. The result is that the soil becomes transformed into a suspension of clay particles in water; a viscous liquid, with almost total loss of shear strength. Once initiated in a small area, the resulting disturbance to the surrounding soil can spread the process. In one such instance, in Sweden, spoil from foundation excavations for a small barn, which was dumped at the edge of a lake, started a sliding process which caused the liquefaction of over $7km^2$ of land and totally destroyed a number of farms.

FROST SUSCEPTIBLE SOILS

These are discussed in Chapter 6. Special freezing and thawing tests can be carried out to measure frost susceptibility but such soils are usually identified by their plasticity index, as indicated in Table 6.2. In many areas, frost susceptible soils are well-known and local practice has been established to deal with the problem. Usual precautions are to take foundations below the expected maximum depth of frost penetration or to replace material down to that depth by a suitable, well compacted fill. Frost susceptibility probably presents more problems for roads than for buildings.

FILLED GROUND

Filled ground causes problems owing to poor compaction, deleterious material and variability within the fill, as discussed in Chapter 6. Where development has to take place on filled ground, precautions are broadly similar to those described for compressible soils.

Even fills laid in a controlled manner with selected material can give rise to difficulties. At the edge of a filled area the change of consolidation characteristics between the fill and adjacent natural ground causes problems of differential settlement. Within the filled area, consolidation of the fill may be incomplete, creating added settlement problems for structures. Poorly-compacted fill may have an open structure, giving it the properties of a collapsible soil, as described earlier. Many areas of backfill in open-pit mineworkings are built with a network of subsurface drains, sometimes supplemented by pumps, to keep the water-table low. When pumping is stopped, or as drains become blocked, the water-table will rise and initiate the collapse mechanism. Alternatively, a rise in the water-table can cause swelling of some soils. A sudden reduction in groundwater levels, produced by drainage of a fill, can also cause problems. The resulting reduction in pore water pressure causes a corresponding increase in effective, interparticle, stress within the soil skeleton (see Chapter 10). This has the same effect as loading the soil and can lead to settlement. This process can also happen in natural soils, as witnessed in the Lincolnshire fens where drainage over two centuries has led to many metres of settlement.

TROPICAL SOILS

A feature of many tropical soils is that they are residual deposits: they are formed by slow chemical weathering of the parent rock which is partly leached away and partly transformed into clay minerals. This produces the open structure of red coffee soils, common in many parts of the tropics. Because of their structure, these may be prone to collapse on flooding, as discussed earlier in this chapter. Black cotton soil is another tropical soil that can give problems: it is a highly expansive soil and should be avoided whenever possible. If it cannot be avoided then the precautions suggested for expansive soils should be considered. Tropically weathered volcanic ash can also produce highly expansive soil. If weathered in-situ, the resulting soil can appear deceptively good as a foundation material, with a high shear strength and moderate consolidation properties. On wetting, however, both collapse of the open structure and swelling of the clay minerals can occur.

DESERT SOILS

These are soils occurring in arid and semi-arid regions. The problem of expansive soils in such regions was discussed earlier: the collapsible structure of loess has also been mentioned. In addition to these problems, chemical attack of buried concrete or steel can be a problem. Salts which would have been leached away in wetter climates are left in the soil; and even concentrated, in desert climates. Sulphates are common, often in the form of gypsum, and even bands of common salt (sodium chloride) may be present. Special precautions, such as those indicated in Table 5.10, may be needed. Because of the high incidence of aggressive salts in desert areas, special care has to be taken over the selection of aggregates for construction materials.

Chapter 8

MINING AREAS

INTRODUCTION

The site investigation procedures described in Chapter 3 often have to be modified in Coal Measures areas to allow for the unique problems caused by the presence of old mine workings. The desk study assumes greater prominence and the ground investigation entails deep boreholes penetrating much further into the underlying rock than would be the case for a normal site. The implication for the developer is a higher cost site investigation and the possibility that expensive remedial measures will be required to safeguard against mining subsidence. Even if old mine workings are not present, the developer may be faced with the possibility that future coal extraction will produce ground surface movements and resort may have to be made to special foundations or structural frames capable of accommodating differential settlements. This chapter examines the modified desk study and ground investigation procedures required in coal mining areas and also describes the precautions and remedial measures which may have to be taken to avoid distress or failure of proposed or existing constructions.

MINING METHODS

To give an appreciation of the unique problems caused by past mining activity it is necessary to briefly describe early mining methods and to distinguish between partial and total mining techniques.

Early mining took the form of scouring or scraping of coal at exposed seam outcrops and the driving of adits (drifts and levels) short distances into the seam in areas of self-draining high ground. Another general form of early mining was the sinking of shallow "bell pits" just downdip from the seam outcrop, the pit being widened-out (or belled) when the seam was reached. When the widened-out area became unsafe the bell pit was abandoned and another one sunk close to it. This method of mining left a series of shallow pits scattered along the seam outcrop as indicated in Figure 8.1.

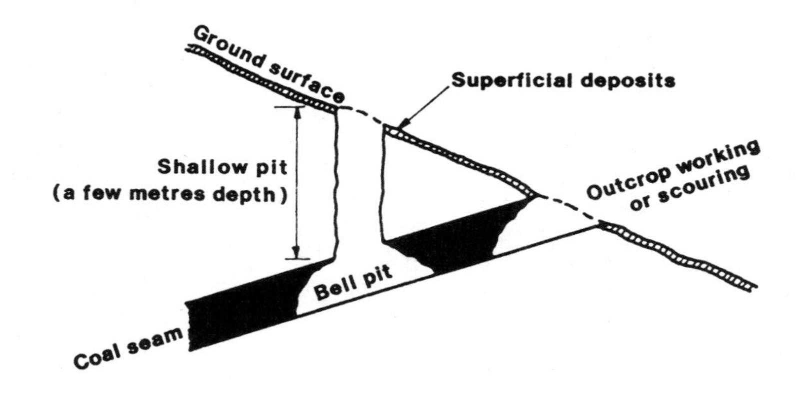

Figure 8.1 Outcrop working and bell pit

Partial extraction mining came into being as soon as workings were taken beyond the few metres depth associated with bell pit mining. The deeper workings went, the higher the pressure imposed upon them by the weight of the overlying strata and, to prevent the workings from being crushed and squeezed out of existence, some of the coal was left in place in the form of large pillars providing resistance to the downward pressure. This technique, variously termed pillar and stall, bord and pillar, post and stall, room and stall, stoop and room, altered in detail between coalfields and even within a single coalfield, but always possessed the common feature of pillars of coal left behind to support the roof, with coal worked from the stalls as illustrated in Figure 8.2. This partial extraction mining is referred to as pillar and stall mining throughout this chapter. A feature of pillar and stall mining was "pillar robbing" in which a second phase of working, usually carried out near the end of a colliery's life, reduced the pillar size or completely removed a percentage of the pillars. This secondary working often brought parts of a colliery to a state of incipient roof failure.

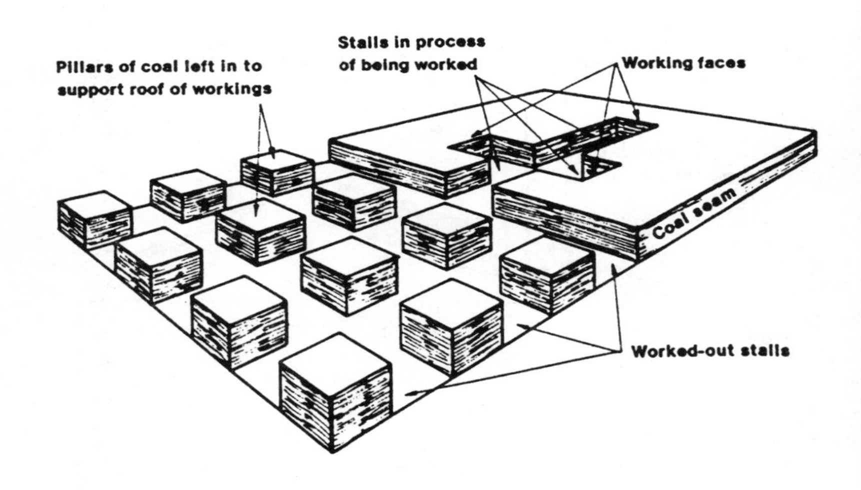

Figure 8.2 Regular pillar and stall working

Total extraction mining also had an early development but did not come into widespread use until the late 19th and early 20th Century. In this method, parallel roadways were driven along the seam and the coal between them totally extracted. The space produced by this working was sometimes partially filled with waste ("stowing") from the mining operations but was usually left unfilled. Withdrawal of roof supports from the worked area produced immediate collapse of the overlying strata with associated ground surface movements taking place at virtually the same time. Because of this contemporaneous nature of the surface movements, it is usually assumed that total extraction or "longwall" mining has left little in the way of problems for present site development.

THE PROBLEMS CAUSED BY THE PRESENCE OF OLD MINE WORKINGS

Old mine workings constitute two main problems to site development - the presence of unstabilised mine entries and the presence of pillar and stall workings at shallow depth below the ground surface.

Mine entries in the form of bell pits, adits and shafts occur in great numbers in all old coal mining areas and have seldom been properly filled-in or capped over. To compound the problem, on abandonment of the colliery, most old shafts were just blocked a short distance from their top, filled over with soil, and hidden from sight. Any structure unknowingly built over an unstabilised entry, even a bell pit a few metres deep, can suffer distress if the blocking or the shaft lining collapses and a void forms under the foundations. The problem is more acute if construction takes place over a large diameter unstabilised shaft which has been concealed under a significant depth of fill. In this case, collapse of the blocking or lining leads to a loss of fill down the shaft with the formation of a crater at the ground surface as shown in Figure 8.3. Any structure founded in such a location would be likely to suffer failure.

Pillar and stall workings at shallow depth. Before examining the problems caused by the presence of shallow pillar and stall workings it is necessary to explain what is meant by "shallow depth". In qualitative terms it can be defined as that depth below the ground surface within which the presence of old pillar and stall workings can have an adverse effect upon the stability of the ground surface or any development at that level. Subsequent discussion will suggest that a rock cover above the old workings of less than 10 times the thickness of the coal seam or height of the workings is an acceptable quantitative

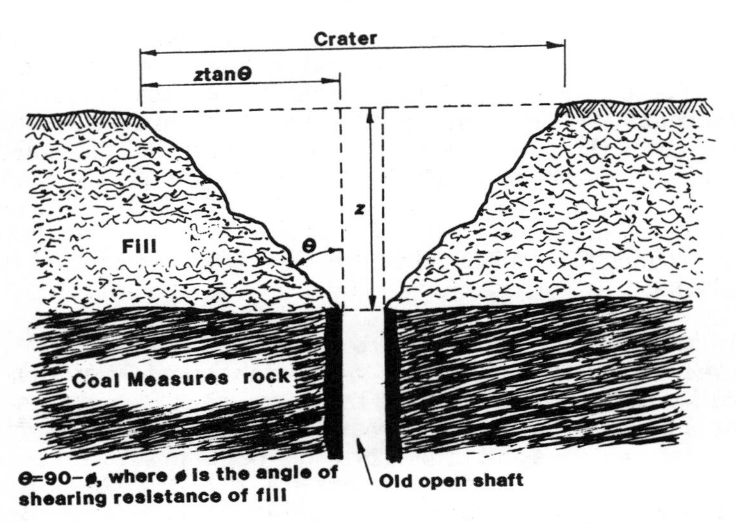

Figure 8.3 Crater formed by shaft collapse below fill

definition of shallow depth.

There are three different failure mechanisms causing the problems associated with pillar and stall workings at shallow depth - void migration, pillar punching and pillar crushing.

Void migration is the most commonly occurring and important of the problems. It takes place when the roof above the stalls and roadways of the old workings becomes incapable of spanning from pillar to pillar, disintegrates and falls into the space below, as illustrated in Figure 8.4. This roof failure takes place progressively with a consequent filling of the original working space by roof debris and an accompanying upward movement of the void; hence the term void migration. The upward movement continues, in a completely unpredictable manner, over a long period of time, finally ceasing when:

(a) its progress is arrested by a competent stratum such as a sandstone layer;
(b) it chokes itself by bulking i.e. the falling roof material, occupying a greater volume than it did in its intact state, catches up with the migrating void and blocks further roof fall; or
(c) it reaches the bedrock surface. When this occurs,

the superficial deposits above the bedrock may collapse into the void with the formation of a "crown hole" at the ground surface. If the superficial deposits are strong and of substantial thickness (e.g. a thick deposit of stiff boulder clay), they will attempt to span the void but will suffer deformation with resulting surface movement, as shown in Figure 8.5. Structures founded in areas where void migration reaches the bedrock surface will inevitably suffer distress.

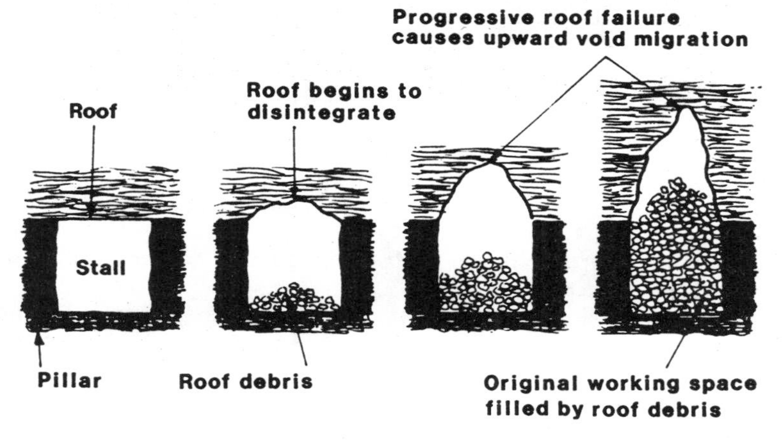

Figure 8.4 Development of void migration

Void migration usually takes place to a height of between 3 and 5 times the thickness of the coal seam or worked space but has been known to reach a height of 10 times that thickness. Because of this, shallow workings are taken to be those lying within a depth below rock surface defined by 10 times the thickness of the extracted seam or working space, whichever is the greater. The height to which void migration will occur under a specific site depends on a number of factors, the most important of which are: the dimensions of the original workings; the lithology (the nature and properties) of the rocks overlying the workings; and the bulking factor of the rocks overlying the workings.

Pillar punching is not as common an occurrence as void migration and is usually associated with the presence of a thick bed of "seatearth" (a fossil

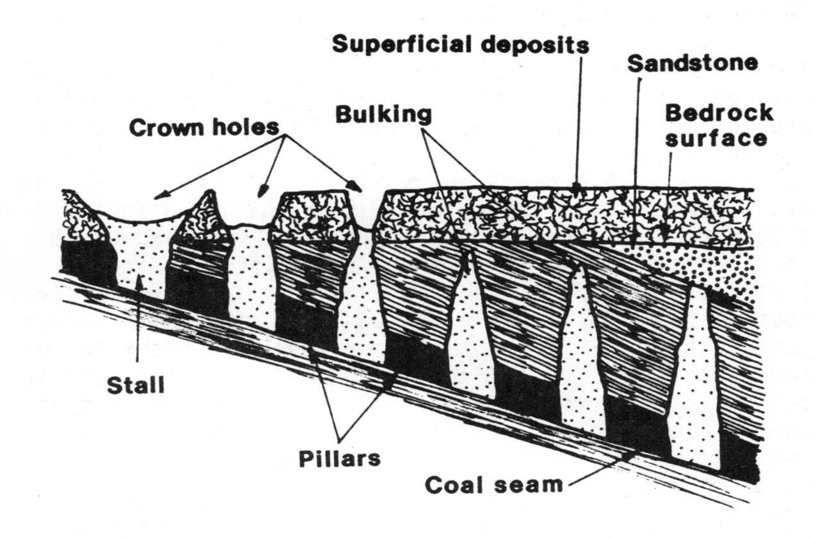

Figure 8.5 Formation of crown holes and arrest of void migration

clay-type soil also termed fireclay because it makes good refractory bricks) under or above the coal seam. The pressure of the overlying strata sometimes causes the pillars to punch through the seatearth floor of the workings or the roof to punch around the pillars. The consequence of pillar punching is a general settlement of the overlying strata producing localised depressions or troughs at ground surface level.

Pillar crushing is an infrequent occurrence at shallow depth and occurs when spalling and weathering, particularly when water is present, causes deterioration of the coal pillars. The decreased strength and bearing area leads to an inability to resist the pressures imposed by the overlying strata and crushing of the pillars takes place. Crushing in one area leads to a transfer of load to adjacent pillars and a widepread failure mechanism can result. The consequence of pillar crushing is a trough-like subsidence at ground level similar to that resulting from modern total extraction longwall mining.

THE ENHANCED DESK STUDY

The desk study assumes great importance for developments in coal mining areas because of the unique problems arising from the presence of mine entries and shallow workings. It is impossible to detect the full extent of these hazards by digging trial pits and drilling boreholes alone, and the only hope of obtaining a full picture of their presence lies in a careful, thorough documentary search backed up by a judiciously planned ground investigation. The extra cost of this greatly enhanced desk study is always justified in terms of both the reduction in the number and depth of expensive rotary boreholes required and in satisfying the professional legal requirement that all accessible relevant information should be consulted. The enhanced desk study involves some procedures common to a normal desk study but, in addition, requires examination of specialised coal mining information held in a variety of sources, particularly plans of workings of abandoned mines,

To properly understand the procedure for carrying out the enhanced desk study it is first necessary to gain an appreciation of the sources and categories of coal mining information available.

Sources of information on past coal mining

These fall into three main groups - Primary, Archive and Other sources, as indicated in Figure 8.6. The figure and the guidance given below refer specifically to Britain but similar overall procedures will be applicable elsewhere.

Primary sources contain information which must always be referred to in the desk study - the abandonment plans of mines originally deposited with the Mines Inspectorate but now in the custodianship of British Coal; any other plans of abandoned mines held by British Coal and included in its schedules and listings; sequential editions of Ordnance and Geological Survey maps; Geological Memoirs; aerial photographs (when available) and any other readily available information. Differentiation has been made between British Coal's holdings of abandonment and other plans. The abandonment plans were originally

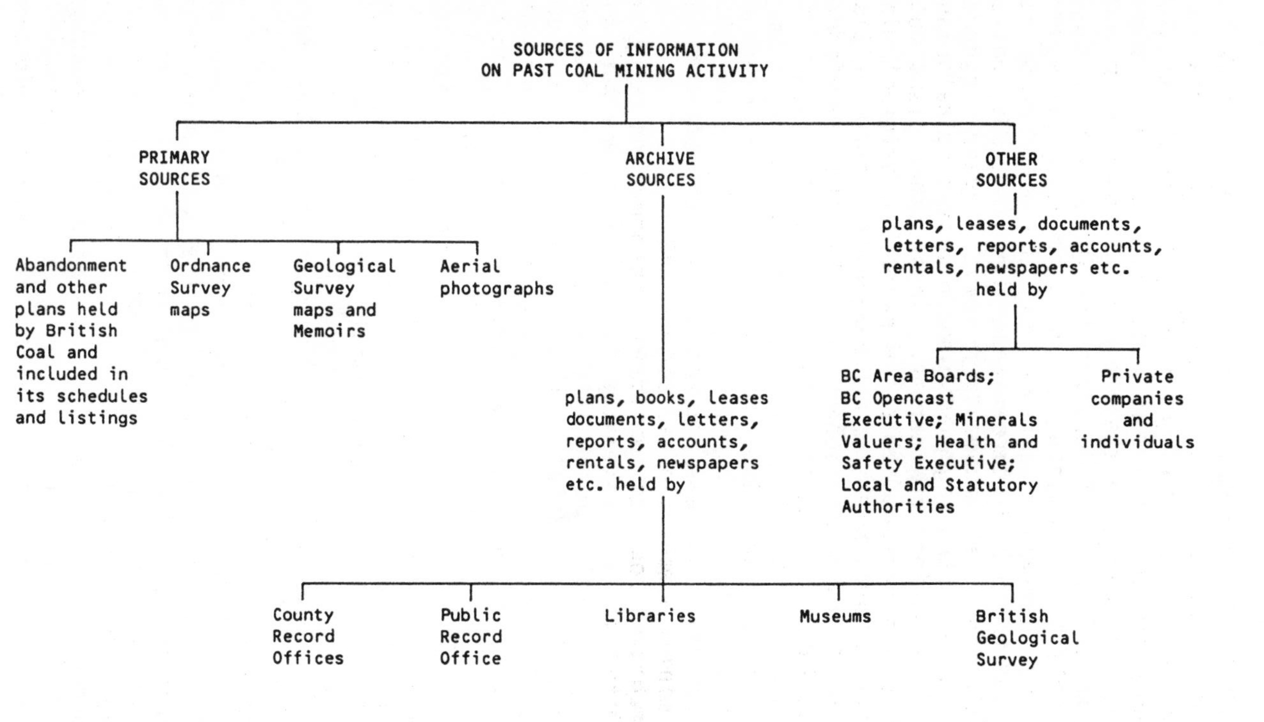

Figure 8.6 Sources of information on past coal mining activity

deposited with the Mines Inspectorate as a result of the Coal Mines Regulation Act of 1872 which made the deposition of working plans mandatory on the abandonment of a colliery. Virtually all these plans postdate 1872. They passed into the custodianship of British Coal in 1950 and are all fully scheduled and available for public inspection. The other plans were not deposited under the 1872 and subsequent Acts and consist of original mine working plans that have passed into British Coal's possession over the years together with copies obtained by British Coal of original plans held in a variety of sources. These other plans both predate and postdate 1872. Many of the original mine working plans have been incorporated into British Coal's Abandonment Schedules with the remainder, particularly the copied plans, listed separately. This situation is changing from year to year and there may be no distinction between scheduled and listed plans in the future. Reference to Primary sources alone will produce an incomplete picture of past mining for any coal mining area exploited before 1872.

Archive sources contain information seldom consulted in any systematic fashion although they are usually scheduled and available for public inspection. County Record offices; the Public Record Office; national, local, university and college libraries; national and local museums; British Geological Survey offices (particularly Nottingham), all hold plans, books, leases, documents, letters, reports, accounts, rentals, newspapers etc. containing valuable information on past coal mining. It has to be acknowledged that information in Archive sources is difficult to consult, assess and use within the time-scale associated with the desk study stage of a site investigation.

Other sources hold the same kind of information as the Archive sources but are seldom scheduled and sometimes not available for inspection. They are found at numerous locations - Area Board offices of British Coal; Opencast Executive of British Coal; Minerals Valuers offices of the Valuation Department of the Board of Inland Revenue; Health and Safety Executive; County and Local Authorities; Statutory or Public Authorities; private companies; private individuals. In a similar manner to the Archive sources, the information held in Other sources is difficult to consult, assess and use within the limited time-scale normally available for a desk study.

Categories of information

In the context of engineering planning and design, the information obtained from the three sources falls into two distinct categories - direct and indirect information.

Direct information allows mine entries to be located and the extent of shallow workings to be plotted onto modern Ordnance Survey or Geological Survey maps. All Primary sources and many of the Archive and Other sources contain direct information, mostly in the form of mine working plans. Direct information is relatively easy to abstract and is immediately understandable.

Indirect information proves or implies the existence of mine entries and shallow workings but does not give their precise location. This information is usually contained in the books, leases, documents, letters, reports, accounts etc. of the Archive and Other sources. Indirect information can be difficult to abstract and needs considerable interpretation before it can be applied.

PROCEDURE FOR CARRYING OUT THE ENHANCED DESK STUDY

Because the desk study assumes such importance when investigating sites in old coal mining areas, the procedures which should be adopted are described in full detail. There are two alternative procedures for the enhanced desk study - a full or "comprehensive" procedure and a shortened or "streamlined" procedure.

The comprehensive procedure is the ideal case in which time and cost are of secondary importance to the achievement of a perfect study. All known sources are consulted and all relevant information obtained. The procedure, shown in chart form in Figure 8.7, is described below.

1. Define the proposed site on modern Ordnance Survey maps at a scale of 1:10560 or 1:10000. This is a suitable scale for valid initial

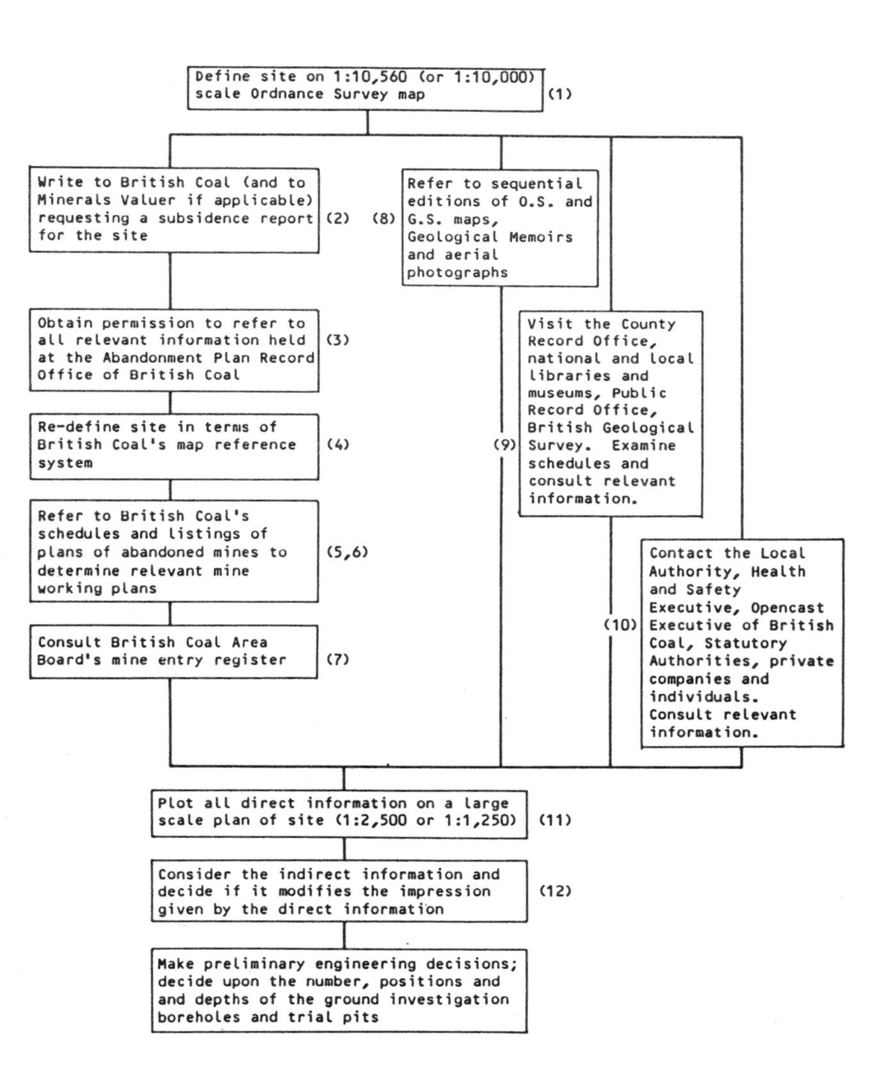

Figure 8.7 Comprehensive procedure for desk study in coal mining areas

impressions of the site; is identical to the scale of the detailed Geological Survey maps which have to be referred to at all stages of the desk study; and is a scale which has to be used to define the site in step 4 below.

2. Write to the Area Surveyor and Minerals Manager of the relevant Area Board of British Coal (and to the relevant Minerals Valuation office of the Inland Revenue if representing a Local Authority, Government Department, Statutory Authority or other approved Governmental body) requesting a subsidence report on the proposed site. This will provide a general statement of subsidence risk due to past, present and future mining. The report, although informative and providing guidance, is not sufficiently detailed for decision-making processes and usually errs on the side of caution.
3. Write to the Area Surveyor and Minerals Manager and arrange to consult the Abandonment and other plans and any other information scheduled and listed at the relevant Abandonment Plan Record office of British Coal.
4. Re-define the proposed site in terms of the grid reference system adopted by British Coal. This will be the starting point for the search for relevant mine working plans. This re-definition can be carried out in advance of the first visit to the Abandonment Plan Record office if you have knowledge of the system, but is usually done with the help of British Coal staff during the first visit.
5. Refer to British Coal's schedules (also termed catalogues)and listings of plans of abandoned mines to determine which show workings under the grid reference squares covered by the proposed site.
6. Order copies of relevant plans in order to allow working and subsequent checking to be carried out conveniently back at your office. Plans which are too large, too fragile or unsuitable for copying will have to be worked from at the Abandonment Plan Record office.
7. Refer to British Coal's mine entry register if permission can be obtained. This is a register

of all mine entries known to British Coal, with entry locations plotted on 1:2500 scale Ordnance Survey Maps: British Coal has examined every plan or copy plan in its possession for evidence of shafts and adits. The mine entry register provides the most complete and up-to-date record of known shafts and adits but is often not available for consultation by the public. British Coal maintains that, because of the inaccurate nature of many old mining plans, entry location is necessarily approximate and responsibility for accurate entry location must lie with the developer. This is true, but many thousands of hours of work have been put into the compilation of mine entry registers and a source of such valuable information should be available to all interested parties. Besides containing comprehensive information on all known mine entries, the register also provides valuable clues to the colliery and seam names which the searcher needs to know when scanning the schedules and listings for relevant plans. A request to consult the mine entry register should be made in every instance.

8. Refer to sequential editions of Ordnance Survey maps at scales of 1:10560, 1:10000 and 1:2500 and Geological Survey maps at scales of 1:63360, 1:50000, 1:10560 and 1:10000 if available (County Record offices, libraries and museums hold old editions). Consult the Geological Memoirs accompanying the Geological Survey maps. Aerial photographs can be referred to, if available, but may not provide additional information to that already shown on the Ordnance Survey and Geological Survey maps.
9. Visit the County Record office and examine schedules of deposited documents of estates and companies which may have been involved with coal mining in the region of the site. The archivists will provide guidance but you will have to be specific in terms of colliery names, estates, companies and the entrepreneurs involved. Repeat the process at libraries and museums. The local library may also hold books relating to mining in the region of the proposed

site. The London and Nottingham offices of the British Geological Survey, the relevant regional office of the British Geological Survey, and the Public Record Office, Kew, may also hold information. It has to be acknowledged that this search of Archive sources is difficult, time consuming and expensive, with relatively little return for the effort put in compared with that obtained from steps 1 to 8 above. Nevertheless, valuable information, unobtainable elsewhere, is contained in these Archive sources.

10. Contact the Local Authority and arrange to meet the engineers and planners to discuss the site: refer to any information they may hold. Write to the Mining Records office of the Health and Safety Executive, London, to check for old fireclay or other mineral working plans and records which often include details of coal mining. Write to the Opencast Executive of British Coal to ascertain whether detailed information (particularly borehole records) is held for the proposed site. As a result of impressions gained and suggestions made at all stages of the desk study, contact Statutory Authorities, private companies and individuals who are thought to hold information on past coal mining activity in the region of the proposed site.
11. Plot all the direct information obtained relating to mine entries and extent of workings on to a large scale site plan - 1:2500 or 1:1250.
12. Consider the evidence of the indirect information and decide whether it is likely to contradict or modify the impression given by the direct information e.g. the direct information may show workings at significant depth only whereas the indirect information may suggest or even prove that shallow working took place.
13. The final picture provided by this comprehensive procedure allows preliminary engineering decisions to be reached and enables an efficient and economic ground investigation to be planned in terms of the number, position and depth of

boreholes and trial pits. Correction of some seam outcrop positions and fault representations shown on the Geological Survey maps may also prove possible.

The streamlined procedure is resorted to in most cases because the time and money available for the desk study are usually insufficient to allow the comprehensive procedure to be followed. It is important that the optimum return, in terms of the presence of mine entries and shallow workings, be obtained and a selective procedure concentrating on the direct plan information in the Primary and Other sources helps achieve this. The streamlined procedure, shown in chart form in Figure 8.8, is described below.

1. Carry out steps 1 to 7 of the comprehensive procedure i.e. request subsidence reports and refer to all British Coal's direct information held in plan form. In doing this, much of the direct plan information held in Archive and Other sources will be consulted as British Coal holds copies of many of these plans. The efficiency of this operation depends heavily upon the extent to which the Area Board has collected plans from its regional offices and collieries and has obtained copies of plans held by Record offices, libraries, museums and other organisations. The situation varies between the various Area Boards of British Coal and even for different regions within a given Area Board.
2. Carry out step 8 of the comprehensive procedure, omitting examination of aerial photographs if they are not readily available.
3. Carry out the first part of step 10 of the comprehensive procedure i.e. contact the Local Authority. Planning permission for the proposed construction is required from the Local Authority in most cases, anyway, and they, together with British Coal are the people most likely to provide valuable guidance and advice.

This shortened or streamlined procedure can usually be carried out within the time-scale normally associated with a site investigation and can provide a reasonable compromise between the comprehensive search of all information sources and a limited search of readily available Primary sources alone. A limitation of this procedure is the absence of the indirect

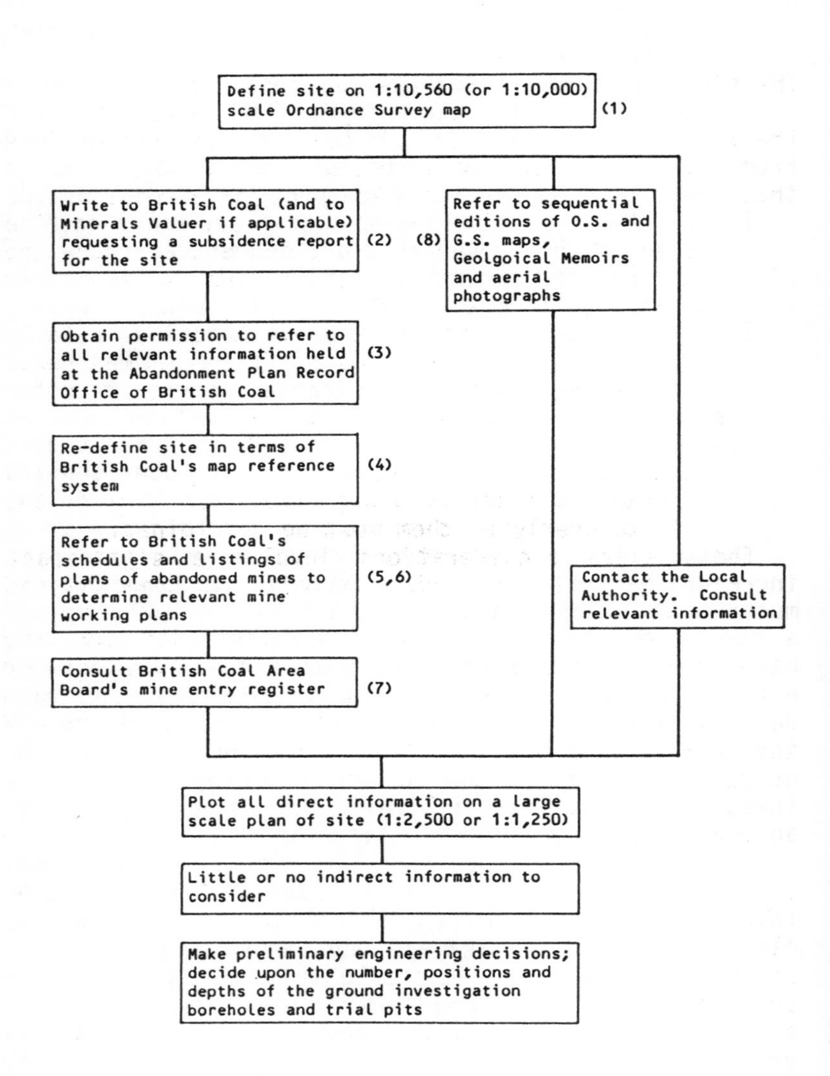

Figure 8.8 Streamlined procedure for desk study in coal mining areas

information which so often helps influence the concept of the proposed development.

THE MODIFIED GROUND INVESTIGATION

The ground investigation in a coal mining area differs from that described for a normal site in Chapter 3 in that three extra major considerations have to be made:

1. all mine entries known to be present from the desk study, but not visible on site, must be accurately located in order that they can be stabilised;
2. the coal seam outcrop positions, dip representations and number of coal seams shown on the geological maps must be verified in order that zones of possible shallow workings can be correctly identified; and
3. the presence and condition of old shallow workings and the void migration potential of the strata overlying them must be determined.

These extra considerations involve a significant increase in the number of trial pits or trenches and much deeper boreholes than required for a site outside a coal mining area. Geophysical techniques may also have to be employed in an attempt to locate concealed entries and the presence of shallow workings. A good desk study will reduce the size and cost of the ground investigation and facilitate the interpretation of the borehole records, but a more expensive ground investigation is inevitable when dealing with a site in an old coal mining area.

Location of mine entries will have already been carried out at the desk study stage of the investigation by plotting shaft and adit positions from old mining plans onto modern Ordnance Survey maps at a scale of 1:2500 or 1:1250. Site reconnaissance can sometimes confirm the exact location of the plotted entries, but, all too often, there is no trace at ground surface level of the shafts and adits involved. Additionally, because of the inaccuracies inherent in the majority of old mining plans, particularly in terms of ground surface features, the "location" obtained from the desk study may really be a zone, perhaps of the order 30 metres square. The ground investigation in these cases involves searching a substantial area of the site as distinct from confirming a precise location.

Trial pits, trenches and boreholes are usually employed to confirm the position of, or to search for, a concealed entry. If the site has not been filled over to any significant depth it is normal practice to consider trial pits and trenching for the first attempt. These are usually patterned in a systematic manner over the zone indicated by the desk study. If the depth of filling is great, or if the water table is high, boreholes may have to be employed. It is normal practice to start at the centre of the most likely location given by the desk study and to adopt an outward spiral sequence of subsequent boreholes based on a "square search" pattern until the entry is located, in the manner suggested in Figure 8.9. If the entry cannot be located by these methods, resort may be made to the geophysical techniques described in Chapter 3, although their success rate has not been high in the past. One of the reasons for this lack of success is the presence of so many subsurface remains at old colliery sites, all of which are likely to show up as ground anomolies similar to those of concealed entries.

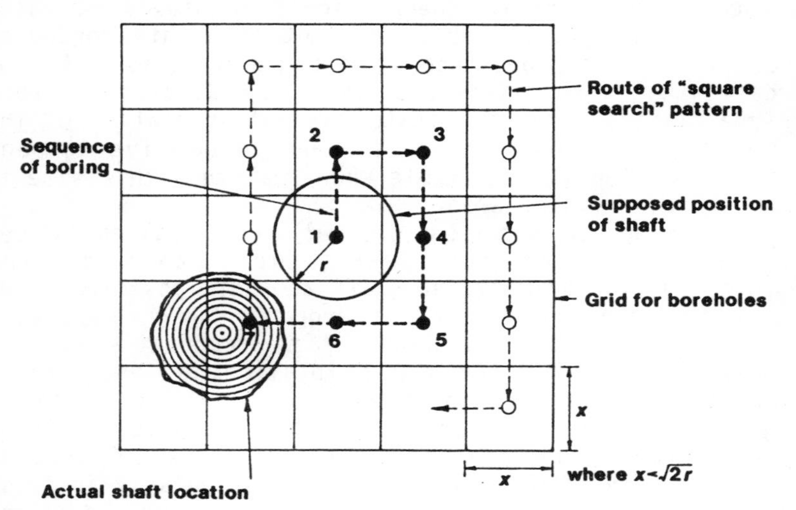

Figure 8.9 Square search borehole pattern for location of a concealed shaft

If these methods fail to reveal the entry, large scale excavation may have to be employed over the entire area to be occupied by the proposed construction. This solution is sometimes advocated for sites of limited area which have not been filled over to any significant extent; the presence of all entries, those known from the desk study and those previously unsuspected, would be revealed by stripping off the topsoil and fill.

Safety precautions must be taken when searching for old mine entries. The blocking to a shaft can suddenly collapse, particularly when the fill above it is being disturbed or subjected to pressure, and it is important that personnel and equipment are adequately protected. Protective measures can range from safety harnesses for workmen through to support of drilling rigs or excavation plant on grillages spanning beyond the area being investigated. A re-usable textile mesh pegged into the ground over the zone of the site being explored can also provide a safeguard against the possibility of personnel or plant falling down a suddenly collapsing shaft.

Verification of coal seam outcrop positions shown on the 1:10560 or 1:10000 scale geological map(s) covering the site is an essential consideration because all estimates of possible shallow working hazards relate to seam outcrop. Examination of any geological map reveals that the vast majority of seam outcrop positions shown are conjectural (represented by dashed lines as distinct from the full lines which indicate certainty) and any estimate of shallow workings based on such representations will not necessarily be correct. Site reconnaissance can sometimes confirm the accuracy of the representation shown on the geological map, but this is not a common occurrence, being possible only when the Coal Measures rock is exposed or when lines of old shallow workings follow the outcrop. All too often evidence of the seam outcrop is totally concealed under a featureless mantle of superficial deposits and resort has to be made to digging trial pits and trenches or drilling boreholes to verify the position of the seam.

If the depth of superficial deposits overlying the Coal Measures rock is not too great (up to a few metres), and if the water table is not too high, an excavator can be used to dig trenches aross the seam outcrop positions shown on the geological map. Alternatively, lines of trial pits can be dug, but this takes much longer and, if the actual outcrop lies between pits, it may be missed. Trial pits are regularly used, however, because they can form part of the ground investigation programme with the taking of samples in the superficial deposits.

If the depth of superficial deposits is too great, or the water table too high, to allow trial pits and trenching, lines of boreholes have to be sunk across the seam outcrop position shown on the geological map. These boreholes may be shallow, sunk solely for the purpose of proving the seam outcrop, in which case they will be closely spaced in straight lines from updip to a short distance downdip of the shown outcrop position. They will penetrate the Coal Measures rock only to a depth sufficient to prove the seam or old workings in the seam. More usually, though, the boreholes will form part of the main ground investigation with the taking and logging of rock cores and will be drilled to the order of depth necessary to prove the positions of all seams likely to affect the site. These boreholes will be more widely spaced and the seam outcrop positions will have to be estimated from the geological sections constructed from the borehole logs. To ensure accurate sections it is important that boreholes be laid out in straight lines in the direction of the dip of the seam.

Verification of the dip of the coal seams is also an important consideration in the assessment of zones of shallow workings. Geological maps do not generally show a large number of dip values and it is often the case that no dip arrow appears in the region of the site considered. The mining plans consulted during the desk study will have provided dip values, but when these plans were prepared, they often gave the average slope encountered in the workings, usually presented in an easy to understand, rounded, form based on linear measurement, e.g. 12 inches in the yard, 1:4. When workings approached a fault and the dip of the strata altered rapidly in magnitude and direction, this was seldom recorded. Because of these factors, it is important that dip angles be carefully calculated from the geological sections derived from the borehole records, in order that variations in depth of shallow workings can be accurately assessed.

The number of coal seams shown on the geological maps should also be carefully appraised in terms of the findings of the borehole records in particular, and possibly in terms of the findings of the trial pits or trenches dug for the purpose of verifying seam outcrop positions. Seams not shown on the geological maps may be present and shallow workings could exist. A good desk study should have provided evidence of such workings, but careful study of the ground investigation records could also confirm the presence of unrepresented seams. It is important that all anomolies in coal seam occurrences in the borehole records be appraised; a thin seam, uneconomic to work at depth, could have been exploited at shallow depth in early times and small voids can be as damaging to structures as large voids.

The presence and condition of shallow workings should be assessed. Once zones of potential shallow workings have been identified it is important that an attempt be made to physically assess the extent of the workings, whether pillar robbing has taken place, the height to which void migration has occurred and the general state of the strata overlying the old workings. Occasionally a situation exists, usually in naturally draining high ground areas, where the old workings can be entered and their condition assessed. In most cases, boreholes will have to be drilled and the presence and condition of the workings assessed from the borehole logs and rock cores.

The depth and spacing of boreholes must be related to site conditions and to the purpose for which they are drilled. It has been seen that boreholes into the Coal Measures rock are needed for verification of seam outcrop positions, seam dips, numbers of seams and for the presence and condition of shallow workings.

The depth below the rockhead to which the boreholes should be taken varies with their purpose. Boreholes drilled for the dual purpose of defining the geology of the site and assessing the presence and condition of shallow workings should be taken to depths which intersect all the coal seams influencing the site. These boreholes may have to penetrate as much as 50 to 60m below rockhead, but this would depend upon the size of the site being investigated. Boreholes drilled solely for assessing the presence and condition of known or suspected shallow workings should be taken to below the coal seam in question. Irrespective of purpose, all boreholes should be carefully logged and all encountered voids or zones of incomplete core recovery should be noted. The spacing and positioning of boreholes depends upon details of the site and the proposed construction, but every attempt should be made to obtain comprehensive geological sections, particularly in the direction of the true dip of the strata. Boreholes are often based on a regular grid aligned in the assumed dip and strike directions of the strata for this purpose. Boreholes sunk for assessment of shallow workings should avoid a regular grid pattern to prevent the possibility of the adopted spacing coinciding with the coal pillars of the uniform pillar-and-stall-type workings shown in Figure 8.2. If this occurred, it might be concluded that the seam was unworked.

Geophysical techniques of the type described in Chapter 3 do not, in their present state of development, appear to be suitable for assessment of the presence and condition of old shallow workings.

PRECAUTIONARY AND REMEDIAL MEASURES FOR CONSTRUCTION IN OLD COAL MINING AREAS

As already seen, the presence of unstabilised mine entries and shallow pillar-and-stall-type workings must be assessed when construction is proposed in old coal mining areas. The site investigation should reveal the extent of these hazards and, on the basis of its findings, decisions will have to be taken on the precautionary and remedial measures needed to ensure a safe and economical development of the site. Six possible measures are invariably considered, sometimes singly but more usually in combination with one another, as soon as the findings of the desk study are known, with final decisions reached when the ground investigation is completed.

These measures are:

1. the filling and capping of all mine entries present on the site;
2. the founding of structures on conventional

foundations in zones of the site free from shallow workings;
3. the founding of structures on rigid or raft foundations in some zones of the site underlain by shallow workings;
4. the stabilisation of shallow workings by grouting;
5. the removal of shallow workings by bulk excavation; and
6. the transfer of structural loading to below the level of shallow workings by the use of piled foundations or piers.

Measures 1 and 2 are considered for virtually all sites; measures 3 and 4 are regularly considered; measures 5 and 6 are occasionally considered.

The filling and capping of mine entries is really a mandatory measure for all developments because of the safety risk posed to people by the presence of unstabilised entries. Before any work can proceed, it is necessary to consult with British Coal because ownership of virtually all entries and mine workings is vested with that body. British Coal will impose high standards for the filling, plugging and capping of shafts and adits and high costs can be involved. Besides their potential void hazard, shafts and adits can also pose a gas hazard: emissions of methane, carbon dioxide and nitrogen may occur. Methane, which is highly inflammable, presents a fire risk, whilst carbon dioxide and nitrogen, though non-flammable, can lead to suffocation. Ventilation or gas-release systems may have to be provided. This gas emission hazard, combined with the possibility that stabilised shafts can still partially collapse, leads to the general recommendation that structures should not be sited over old shafts even when they have been filled and capped.

Shafts should be completely filled down to their lowest level whenever possible. The filling for the bottom section should consist of coarse hardcore or boulders (not exceeding 300mm diameter) to allow free percolation of mine water, with subsequent filling being clean granular material free from organic content. If blockages are present in the shaft, attempts should be made to remove them but, if these fail, a free-flowing fill such as pea gravel should be employed. This will pass easily through quite small openings without clogging and can allow complete filling of the shaft to take place. If this measure proves unsuccessful, total grouting of the shaft may have to be considered. In some cases, it may be decided to form a concrete or grouted plug near the top of the shaft, extending down to a depth of between one and three times the shaft diameter, either in place of filling or as a safety measure if the underlying filling is considered suspect and likely to settle.

Once the shaft has been filled it should be covered over with a reinforced concrete cap, of width at least two times the internal diameter of the shaft, to safeguard against distress should the walling or lining to the shaft collapse or the fill settle down. The cap should be founded on rock whenever possible but, in many cases, will have to be founded in the superficial deposits overlying the rock. In this event, the excavation should be taken down to a strong deposit with the shaft walling or lining progressively removed as excavation proceeds. The centre of the shaft should be marked on the cap and, if backfill is subsequently placed above it, a corresponding reference peg or marker should be installed at ground surface level. A typical shaft capping is illustrated in Figure 8.10. Vent pipes will have to be provided if a build-up of gas in the shaft filling is considered likely and further special precautions would be needed, both in terms of construction and venting, should a structure be founded above or immediately adjacent to a capped shaft.

Adits, unlike shafts, can extend for considerable lengths under a site and it is often impractical to recommend that no structure be placed above them. Because of this, it is essential that adits be stabilised to a depth sufficient to ensure there is no danger of void migration or strata subsidence above them. If an adit is present at a shallow enough depth under the site, it may prove practical and economic to excavate down into it and place compacted fill back to ground surface level. It is often the case however, that the adit slopes downward quite steeply to significant depth and an open excavation is out of the question. In this event, the adit should be entered and infilled or grouted to a point where the overlying rock cover is considered adequate. The entrance to the

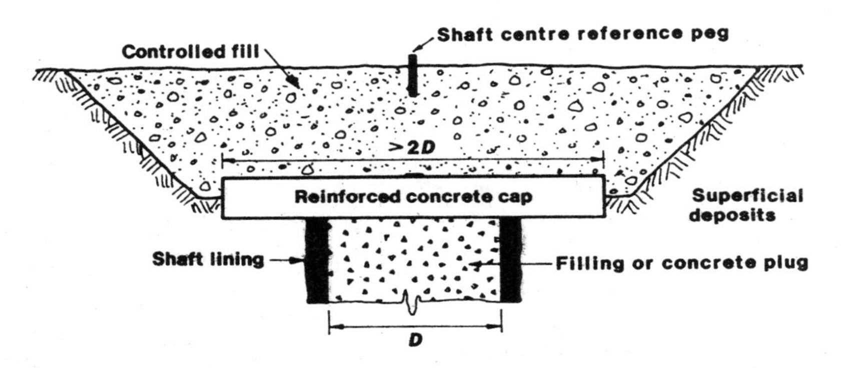

Figure 8.10 Typical shaft capping

adit must be sealed with a brick or concrete walling at least 300mm thick. A typical adit stabilisation is illustrated in Figure 8.11. If an adit has acted as a drainage adit to mine workings in a high ground area, drainage provision will have to be made through the filling or grouting and the walling to prevent a build-up of water in the abandoned workings. Special precautions, such as raft foundations, should be considered for any structure built over a stabilised adit to allow for the possibility that inadequate filling or grouting will allow subsidence to occur.

The founding of structures on conventional foundations in zones free from shallow workings is one of the first considerations made for virtually every site. For each coal seam it is initially assumed that:

1. the zone updip of its outcrop (see Figure 8.12) is free from shallow workings and is safe to build on using conventional foundations; and
2. the zone downdip of its outcrop, where there is a rock cover above old workings of at least 10 times the thickness of the seam or the height of the original workings, is unlikely to be affected by

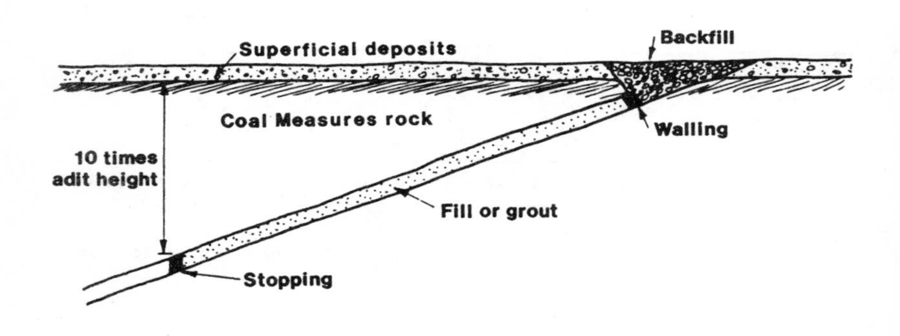

Figure 8.11 Typical adit stabilisation

void migration or surface subsidence and is safe to build on using conventional foundations.

These assumptions lead to two separate "safe" zones, considered free from shallow workings, each side of the seam outcrop as illustrated in Figure 8.12. The delineation of the safe zone downdip of the seam outcrop is usually approximate because the seam thickness, dip and outcrop position on which the estimation is based are seldom accurately known, and even the safe zone updip of the outcrop is dependent upon the accuracy of the outcrop location.
Additionally, there are further considerations which can modify and alter this zoning exercise: the presence of old workings in other seams, close below the one considered, which could propagate voids into the safe zone updip of the seam outcrop or even into the old workings of the seam considered; the presence of a competent sandstone stratum above the seam considered which could arrest void migration and nullify the 10 times seam thickness criterion; the bulking factor of the rocks overlying the seam which, if high, could lead to early choking of the void migration; the possibility that pillar punching or

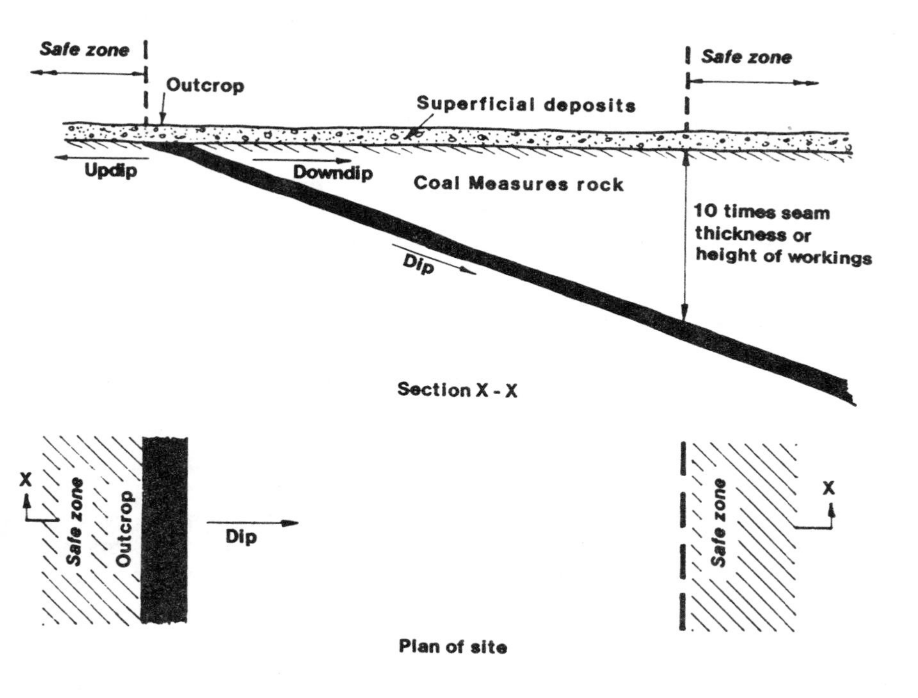

Figure 8.12 Safe zones free from shallow workings

pillar crushing could take place; and the presence of a thick, strong layer of superficial deposits capable of bridging over small crown holes or strata subsidence in the underlying rock.

Despite the many uncertainties always present, it is common practice to allocate safe zones on the basis of the most accurate estimates that can be made for seam thickness, dip and outcrop position, and plan all subsequent development around this approximate analysis. In mitigation, it has to be observed that this is usually the best that can be done because the desk study and ground investigation seldom yield sufficiently accurate information on rock lithology and bulking factor, seam thickness and dip, and the extent of old workings to justify a more sophisticated approach. Nevertheless, when allocating safe zones to a site, particular attention should always be paid to accurate delineation of the seam outcrop, identification of all thick, continuous rock bands above the seam capable of arresting void migration, and assessment of the risk of void migration from lower seams.

The founding of structures on rigid or raft foundations, capable of bridging over small crown holes and minimising the effects of minor surface subsidences, is a measure regularly considered for parts of the site lying between the seam outcrop and the safe zone downdip of the outcrop. It is common practice to assume that those areas where the rock cover above old workings lies between 6 and 10 times the seam thickness or height of original workings is unlikely to be seriously affected by void migration or other subsidence effects, and is safe to build on if rigid or raft foundations are employed. The delineation of this zone is obviously influenced by all the considerations previously listed for the "safe" zones and the final zoning exercise is necessarily approximate.

This assumption of a "raft foundation" zone, combined with the previous assumption of "safe" zones, leads to the implicit conclusion that the area between the seam outcrop and points where the rock cover above old shallow workings is 6 times the seam thickness or height of original workings is unsafe to build on. On this basis, any site can be divided into "safe", "raft foundation" and "unsafe" zones as illustrated in Figure 8.13. The width of these zones, in plan view, can be simply calculated for any site from knowledge of the seam thicknesses or original working heights and seam dips. Alternatively, reference may be made to the type of chart illustrated in Figure 8.14 which gives the relationship between zone widths (in multiples of seam thickness or original working height) and seam dip angle. It should be noted that the rockhead surface has been shown as horizontal in Figures 8.12 and 8.13 whereas, in practice, it may well slope. This must be taken into account when determining the zone widths. Narrower zones result if the rockhead surface slopes upwards downdip of the seam outcrop; wider zones result if the rockhead surface slopes downwards.

It is regularly stressed that the division of sites into safe, raft foundation and unsafe zones is, at best, an approximate exercise and decisions should really be based on a more rational appreciation which takes additional factors into account, such as extent, type and state of old workings, rock lithology and bulking factors. In practice, even with a good site investigation, it is usually difficult to quantify these factors and properly incorporate them into the design with the result that most sites are developed predominantly on the basis of rock cover zoning.

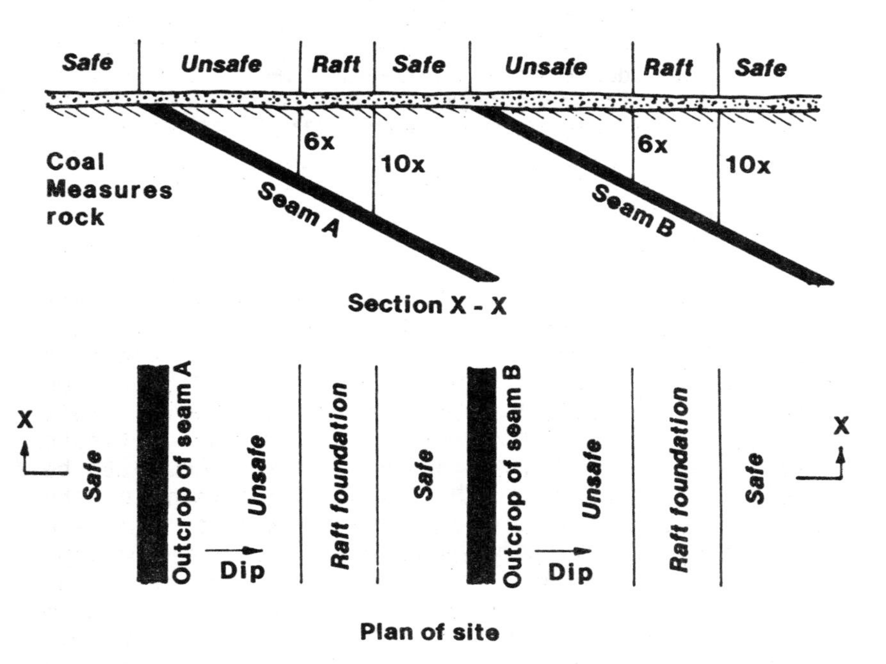

Figure 8.13 Division of a site into safe, raft foundation and unsafe zones

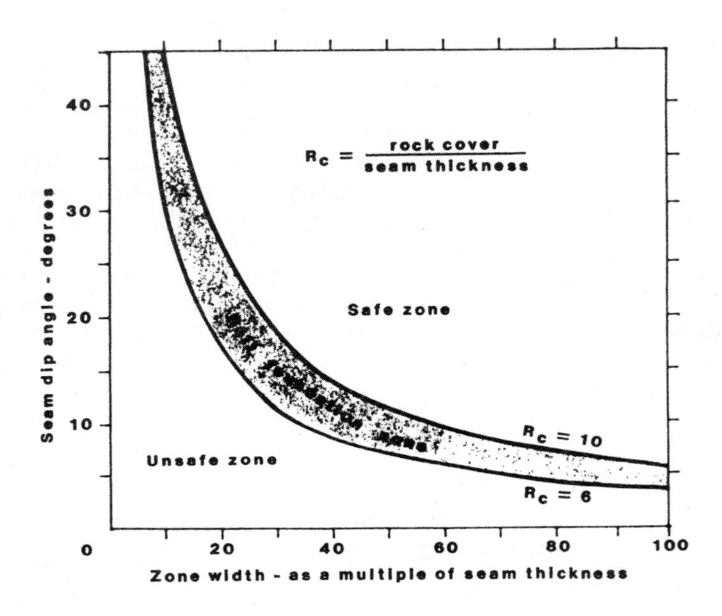

Figure 8.14 Relationship between zone widths and seam dip angle.

The stabilisation of shallow workings by grouting is a regularly considered measure. Injection of grout into the spaces and cavities of old workings, and their associated void migrations and rock fissures, provides roof support, halts void propagation, contributes lateral support to weakened pillars and generally prevents further movements or subsidence from taking place. The benefits to the developer are immense - all problems and uncertainties disappear because approximate zoning solutions do not have to be relied upon and conventional foundations can be employed in all parts of the site, even in the "unsafe" zones immediately downdip of the seam outcrops. Cost is the only drawback; unfortunately, the grouting of shallow workings is an expensive business, often prohibitive in relation to the overall cost of the proposed development, and this precludes its employment in many cases where it could be used to great advantage.

A number of considerations have to be made whenever the decision is taken to grout the shallow workings under a site. The most important of these are: the area over which grouting should be carried out; the depth to which grouting should be taken; and the method and procedure of grouting to be employed.

The area over which grouting is carried out must be sufficient to ensure that no subsidence or crown holes occur under the foundations to the proposed structures. To achieve this aim it is normal practice to grout within an area defined by lines drawn at a given angle to the vertical from the outer limits of the foundations to the bottom of the worked seam as illustrated in Figure 8.15. It can be seen that the

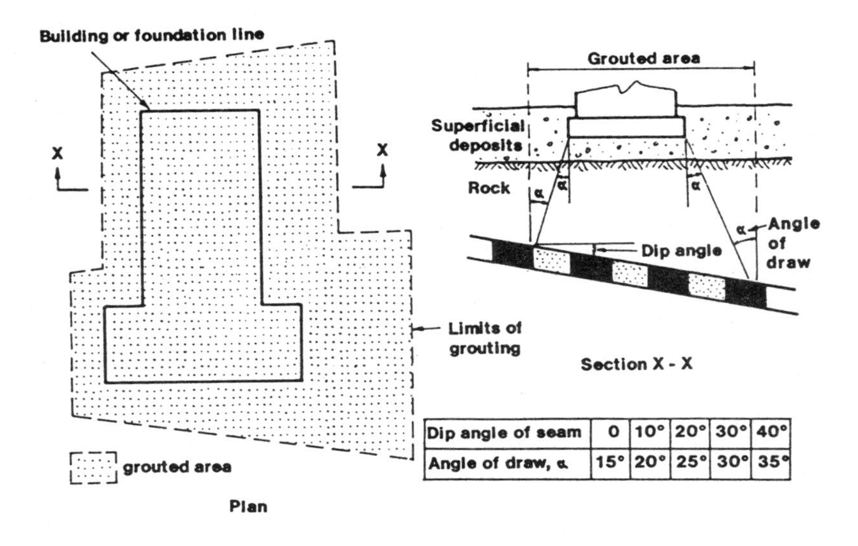

Dip angle of seam	0	10°	20°	30°	40°
Angle of draw, α	15°	20°	25°	30°	35°

Figure 8.15 Area over which grouting is carried out

"angle of draw" can vary between 15° and 35° depending on the dip of the seam, but the upper value of 35° is often adopted, irrespective of the seam's dip. Other empirical rules for definition of the area are also employed e.g. a distance measured laterally outside the structure equal to 0.5 or 0.75 times the depth of the seam below foundation level; a distance measured laterally outside the structure of 10m.

The depth to which grouting should be taken is normally assumed to be the depth to the bottom of the seam whose old workings are considered likely to affect the site. In some cases, workings in lower seams may also be grouted if it is considered that void migration from, or collapse in, them will affect the stability of the upper grouted seam. There are accepted depth limits for grouting, in terms of both physical and economic viability, and it is seldom carried out beyond 50m, with most grouting contracts being for depths considerably less than this.

The method and procedure of grouting depends upon whether or not access can be gained to the old workings and upon the amount of collapse that has taken place within them.

Very occasionally, high level workings may be uncollapsed, dry and accessible through existing adits or by means of shallow excavation or tunnelling. In such cases, perimeter walls, consisting of grout-filled sandbags, are formed around the area to be grouted and a cement/flyash/sand grout pumped through injection pipes, installed at different levels in the walls, until the old workings are totally infilled.

More usually though, workings are inaccessible, partially collapsed and filled with water, and grouting has to be carried out by pressure injection through boreholes drilled from the ground surface to just below the seam in question.

In cases where roof collapse has not occurred to any significant extent, and the working voids have remained open and interconnected, a perimeter wall has to be formed around the area to be grouted to prevent the grout escaping when infilling takes place. The wall is constructed by drilling a large number of closely spaced holes (typically 75 to 100mm diameter at 1.5m centres) and injecting a viscous grout, consisting of various mixes of cement, flyash, sand, gravel and bentonite, through each of them to form a series of interconnected grout cones. When the wall has set, the infill grouting, consisting typically of cement/flyash but sometimes of cement/flyash/sand or gravel, is injected through a regular grid of more widely spaced holes (typically 50 to 75mm diameter at 3 to 6m centres). Injection is continued at each hole until grout escapes from the top of the hole or a pre-determined grouting pressure is reached.

In cases where roof collapse has occurred and the workings have not remained open, consideration should be given to the need for a perimeter wall. It is often decided to proceed directly with the infill grouting and, if the "take" (the amount of grout acceptance by the workings) is not excessive near the boundaries of the area, to dispense with the perimeter wall. Because the voids are not fully open and interconnected, a more fluid grout is often used and a closer pattern of holes (3m centres) adopted. Individual holes, additional to the grid pattern adopted, may have to be drilled if high "takes" suggest the presence of localised large voids.

The removal of shallow workings by bulk excavation is a measure occasionally considered for a limited area development where the workings are close to the ground surface (less than 10m or so) and overlain by weak, collapsible rocks such as shales and weathered mudstones. The allocation of safe and raft foundation zones is always problematical for this type of site and grouting is regularly turned to as a solution. It may be the case, particularly if the site is to be profiled by cut and fill or if deep basements or underground car parks are to be provided, that it would be more economical to excavate down to sound rock under the seam, establish all foundations to the main structure(s) at this level, and backfill the site in a controlled manner with imported material and with any of the excavated soil and rock that proves suitable for the purpose. Any ancillary structures, whose foundations do not bear on the rock below the excavated seam, would have to be founded in the backfill, and, to allow for the settlements that will inevitably occur even in controlled fill, should be provided with rigid, raft foundations. Normal excavation plant would be able to cope with removal of the weak rocks and there could even be some offsetting of costs by sale of the coal obtained from the pillars left in the old workings.

One of the main reasons why this measure is not adopted more regularly is the fact that a straightforward building contract, a high-rise block perhaps, is turned into a major, possibly difficult, earthworks contract. Nevertheless, there are many documented instances of the successful employment of bulk excavation as a method of counteracting the adverse effects of shallow workings.

The transfer of structural loading to below the level of the shallow workings by the use of piled foundations or piers is an occasionally adopted measure, particularly for high rise buildings on sites where old workings are overlain by weak, collapsible rocks such as shales or weathered mudstones. Cast-in-situ, large diameter bored piles are normally employed because they can be augered down through weak rocks without too much difficulty, properly socketed into competent rock under the old workings and, because of their large diameter, are capable of resisting horizontal forces caused by movements in the old workings or overlying strata. They are usually sleeved, partly to ensure that the pile body is correctly formed without too much loss of concrete into surrounding cavities, and partly to eliminate downward forces imposed by settlements and subsidences in the old workings and overlying rocks. Workings in lower seams must be grouted if there is any possibility of void migration or collapse propagating upwards and reaching the pile toes.

Piled foundations are often used in conjunction with the bulk excavation and subsequent backfill measures previously described, and many problem sites could be safely and economically developed by a carefully planned combination of piling and grouting.

PROBLEMS CAUSED BY PRESENT AND FUTURE MINING

Present mining is mostly carried out using a total extraction method, known as longwall mining, in which all the coal is extracted, with the roof of the workings allowed to collapse behind the working face.

This produces a virtually contemporaneous collapse in the overlying strata which manifests itself as a subsidence wave at the ground surface, advancing forwards at the same rate as the workings. This form of subsidence can be particularly damaging because, in addition to causing vertical ground movements, it also produces horizontal ground movements which subject structures to tensile and compressive strains as the wave advances under them. These damaging effects can be minimised if special foundations and structural frames are provided, and any finalised plans to undermine the site must be fully ascertained at the desk study stage of the site investigation.

Future mining is still likely to be carried out mostly by longwall methods, although the contraction of British Coal and the accompanying increase in private mines is already leading to a resurgence of pillar-and-stall-type mining because many of the smaller enterprises cannot afford the high cost of the roof support systems required for longwall extraction. Problems caused by future mining may, therefore, become a combination of both pillar and stall and longwall mining. It is also becoming more difficult to decide whether special provision should be made to accommodate the adverse effects of longwall mining because British Coal is no longer able to be specific about the future exploitation of most coalfields. A developer, faced with the reply that there are no present plans to mine under the proposed site although the right to do so is retained, is unlikely to outlay additional money on special foundations and structural frames.

Longwall working and its effect

Longwall working is usually carried out by driving parallel main roadways along the seam and extracting all the coal between them. The extracted panels of coal may be up to 300m in width as illustrated in Figure 8.16. The roof along the working face is held up by hydraulic supports and, as the coal is extracted and the face advanced, the supports are withdrawn and moved forwards to maintain the new working face region. The unsupported roof behind the working face collapses into the void created by the previous working (the goaf or gob) and the subsidence effect carries through the

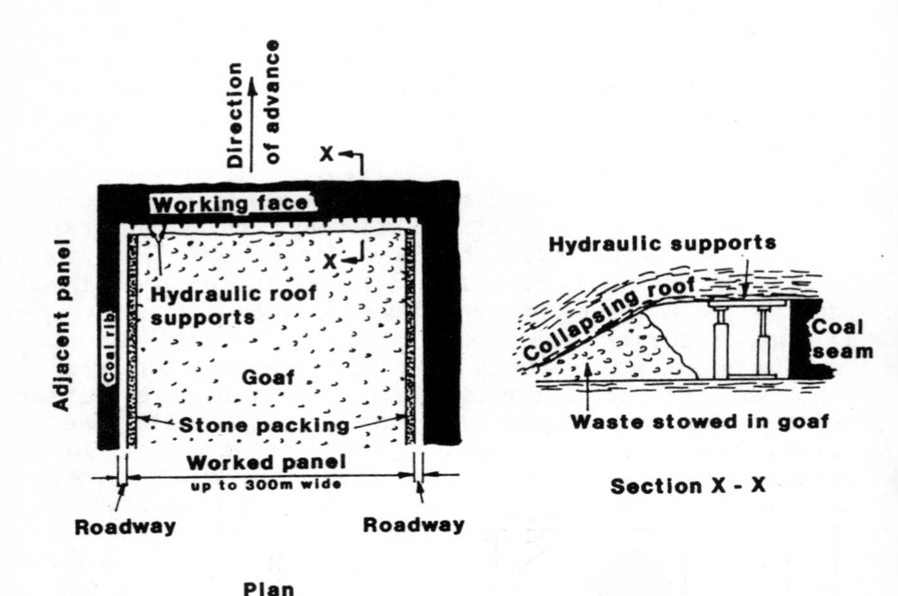

Figure 8.16 Longwall working

overlying strata, finally reaching the ground surface a short time after the working face has been advanced. The roadways are maintained by use of permanent roof supports, by tightly packing waste stone from the mining operations on the goaf side of the roadways and by leaving ribs of intact coal between adjacent panels. In some cases, general waste from the mining operations is stowed into the goaf behind the working region and this helps reduce the amount of subsidence that takes place when the roof supports are withdrawn.

The effects of longwall working are complex and cause ground surface movements above, ahead of, and to the sides of the extracted coal panel. The ground surface area likely to be affected by the workings is defined by means of an angle of draw (also called the limit angle or angle of influence) as illustrated in Figure 8.17. The value of this angle can vary with the dip of the coal seam and the depth and type of

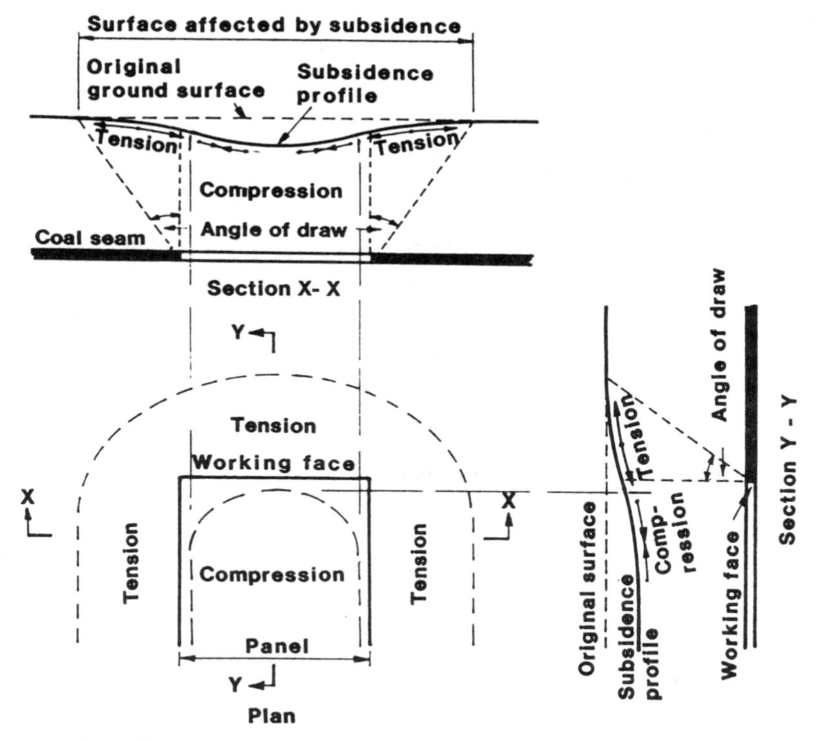

Figure 8.17 The effects of longwall working

overlying deposits, but is usually assumed to be 35°.

Maximum vertical subsidence tends to occur above the centre of the extracted panel and can be as much as 90% of the thickness of the extracted seam in the case of unstowed goaf. Solid stowing of the goaf can halve this value but the costs involved in this operation are high and usually prohibitive. A general relationship between vertical subsidence and width/depth ratio for stowed and unstowed workings is illustrated in

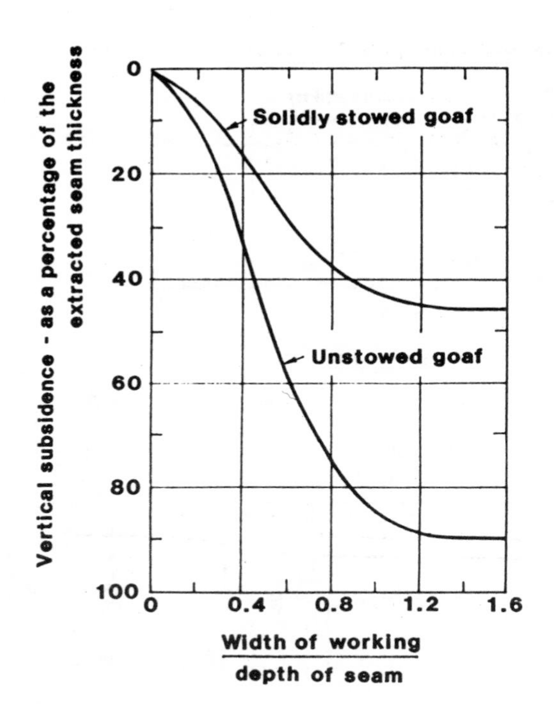

Figure 8.18 Relationship between vertical subsidence and width/depth of workings

Figure 8.18.

Horizontal movements are also present within the subsidence wave and, added to the vertical movements, produce zones of tension and compression at the ground surface as illustrated in Sections XX and YY to Figure 8.17. Structures in the path of the subsidence wave will experience tilting, stretching and crushing and will suffer distress unless special foundations and structural frames have been provided. Figure 8.19 helps explain these mechanisms for a structure directly in the path of the centre of a longwall panel. The structure, a long distance ahead of the working face, is unaffected by ground movement as illustrated in (a).

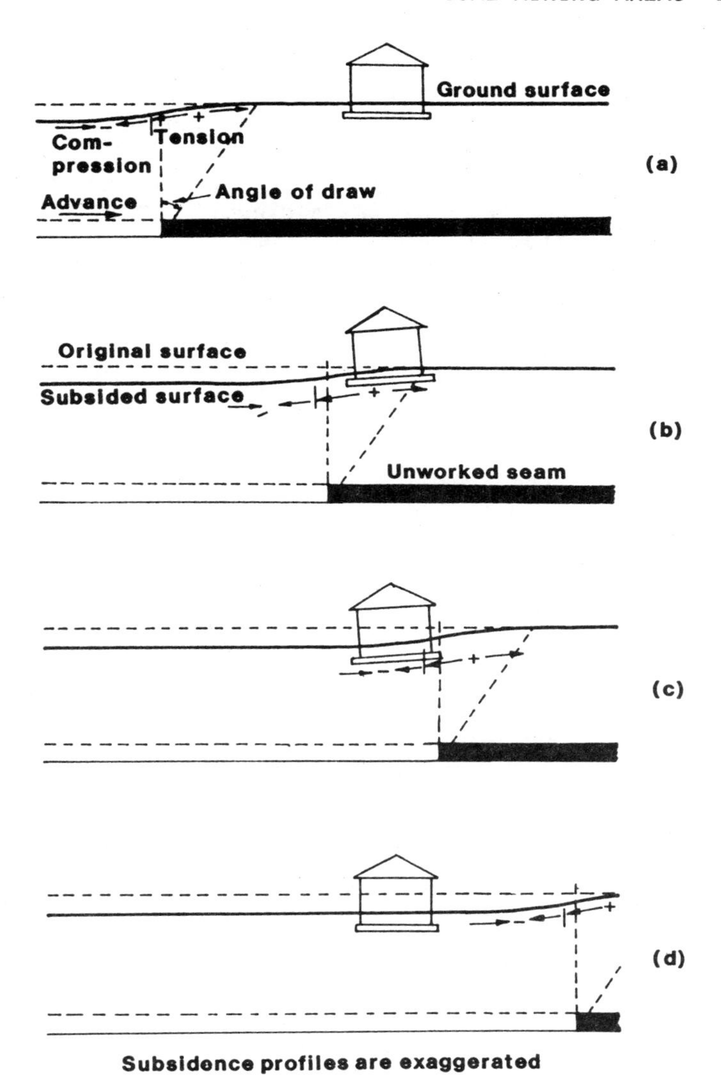

Figure 8.19 The effect of a subsidence wave: (a) structure unaffected by workings; (b) structure tilts, settles and is subjected to tensile strains; (c) structure tilts, settles and is subjected to compressive strains; (d) structure recovers but has been affected by tilt, settlement, and tensile and compressive strains.

As the panel advances, the tension zone ahead of the face intercepts the structure which begins to tilt into the subsidence trough, experiences settlement and is subjected to tensile strains which attempt to stretch or pull it apart as illustrated in (b). As the working face advances further, the compression zone behind the face reaches the structure which experiences more tilt, undergoes further settlement and is subjected to comressive strains which attempt to crush or squeeze it as illustrated in (c). When the working face has passed by, the horizontal strains die away and the structure regains a vertical position within the subsidence trough left by the longwall workings as illustrated in (d). A structure situated towards the edge of the subsidence trough would experience a different sequence of movements and strains and could remain tilted after the working face had passed by. Virtually all these ground movements take place simultaneously with the coal extraction, although some residual movement (as little as 5% of the total in most cases) can occur for a few months after mining has ceased.

The combination of vertical settlement, horizontal movement, tilt, compressive strain and tensile strain can be damaging to all types of structures and five degrees of damage are categorised in practice - very slight, slight, appreciable, severe and very severe. These are defined and explained in British Coal's Subsidence Engineers' Handbook which is the standard reference manual used by practitioners in the mining subsidence field. The Handbook also describes the mainly empirical procedures used for calculating the magnitudes of the various subsidence components. The values obtained by these procedures can be accurate when the coal seam is horizontal and the overlying rock and soil deposits uniform in nature. If the coal seam dips steeply, the overlying deposits are non-uniform, faulting is present or more than one seam has been, or is being, worked in the vicinity, the values obtained will be approximate. Additionally, so many structural factors can influence the manner in which the ground surface movements affect the structure (e.g. size and shape of structure, age and state of repair of structure, foundation type, method of construction) that potential damage can be hard to predict. Despite these many limitations, it is important that all ground movement effects be assessed during the site investigation in order that necessary precautionary and preventive measures may be taken for proposed and existing structures. Because of the uncertainties always present in the assessment, and the serious implications of its findings in terms of planning decisions, design and cost, it is essential that the necessary skill, expertise and experience is available at this stage of the site investigation.

Precautionary and preventive measures can be taken to reduce potential damage to proposed or existing structures likely to be affected by longwall mining. They range from the provision of special foundations and structural frames through to changes in the normal coal mining operations, and can be placed into three general categories: measures for proposed structures; measures for existing structures; and mining operations measures for proposed and existing structures. These measures are often used singly but can be employed in combination with each other.

Measures for proposed structures are mainly precautionary measures incorporated in the foundation and/or structural frame in an attempt to achieve a construction capable of resisting ground movements by virtue of its flexibility. Rigid structures could, in some cases, be designed to do the same job, but their cost is usually far greater and they are seldom considered.

The foundations to proposed structures should, whenever possible, consist of flexible, reinforced concrete rafts with smooth bases founded on a granular bed and membrane at as shallow a depth below the ground surface as possible. Such foundations can only generate low base friction and are able slide on the membrane when horizontal ground movement takes place; as a result, tensile and compressive strains are minimised. If a raft foundation is not suitable for the structure and individual footings have to be used, they should also be smooth-based and laid on a granular bed and membrane with additional reinforcement provided to resist tensile and compressive stresses. They should have the minimum possible bearing area, consistent with the structural load to be carried, in order that bending moments are minimised. If beam and slab or cellular construction has to be used, a smooth base and granular bed and membrane should be provided, and consideration should be given to jacking points for balancing differential settlement. Basements and piled foundations should generally be avoided if horizontal ground movements are anticipated: basement walls would be subjected to high horizontal forces; piles could be sheared and pile caps or beams would be subjected to high tensile stresses.

The frames to proposed structures should be made as flexible as possible by providing articulated joints and by splitting the construction into as many simply supported spans as possible. Large structures should be divided into a number of independent structural units by the provision of continuous construction joints passing through both the frame and the foundation. Rigid frame designs should generally be avoided, although a combination of rigid frames with articulated foundations - the "three point support" system - is used in some European countries. Hydraulic jacks can be incorporated under all structural frames, or under individual items of large machinery, to compensate for subsidence movement.

Cladding and internal finishes should be chosen on the assumption that movement will take place and cracking will occur; flexible joints should be provided in all pipelines and services; structures should not be sited close to or across geological faults or fault zones because ground movements tend to be unpredictable, with high differential values, in such areas.

Recommended procedures for the design and construction of flexible structures, ranging from simple one-storey buildings to bridges, are observed in most countries. The most well-known and used system for buildings in Britain is the Consortium of Local Authorities Programme (CLASP) system which incorporates most of the previously mentioned measures into a spring-braced construction capable of resisting the combination of ground movements arising from longwall coal mining.

Measures for existing structures lying in the path of longwall mining are mainly preventive measures entailing alterations to the structural frame, reinforcement and bracing of the structure, removal of vulnerable components and ground protection around the foundation.

Alteration to the structural frame is a measure to achieve increased flexibility. It may be possible to split a structure into a number smaller units, particularly at connecting corridors or extensions, providing weatherproof, flexible joints at the cuts.

Structures which have low tensile resistance can be reinforced by tie-bolting walls or frames together; similarly, walls can be shored or braced to prevent bulging. Arches can be supported at the centre of their spans, stained glass or decorative windows can be temporarily removed, and large normal windows can be taped in the interests of safety.

Ground protection against compressive strain can be achieved by digging trenches around the structure, down to or below foundation level, and backfilling with a compressible material such as graded boiler clinker. The horizontal compressive ground movement is absorbed by the compressible material, with compressive strain and associated damage reduced by up to 80%.

Mining operation measures for proposed and existing structures can be very effective in reducing or even eliminating ground surface movements caused by longwall extraction. Unfortunately, some of the most effective measures are prohibitive in terms of cost and are seldom implemented. Additionally, the respective developments of the ground surface and the underground colliery are not always compatible operations, although arrangements of mutual benefit to both parties can be achieved if working relationships are established at an early enough stage.

The stowage of the goaf has already been discussed, and Figure 8.18 shows that the vertical subsidence can be reduced by up to 50% if the goaf is solidly packed. This cannot eliminate subsidence entirely though, and the costs of the operation are so high that it is seldom resorted to as a means of reducing ground surface movement.

Leaving a very large coal pillar under a structure can afford complete protection from subsidence, but this measure is only considered when the structure involved is of major importance, particularly in terms of safety to the public, e.g. a dam. The compensation payments for the unworked coal are prohibitive in most cases and, in addition, severe subsidence effects, detrimental to other structures, will occur around the edges of the left pillar.

Working two adjacent panels in the same seam to a pre-determined time lag, and working two or more seams at the same time, have proved successful measures in the reduction of damaging horizontal ground strains. In the former method, termed "stepped face working", the subsidence waves and their travelling strains counteract each other with a consequent reduction in ground surface strain; in the latter method, termed "harmonic extraction", the tensile strain induced by the working face in one seam is partially balanced by the compressive strain induced by the working face in another seam. Both of these strain balancing methods require detailed consultation with mining engineers and geologists for their successful implementation. In practice, there usually has to be a compromise between the maximum possible reduction in ground strain and the problems of economic working at the mine involved.

Working to a combined longwall panel/pillar technique can also produce significant reduction in ground surface movement and is a measure sometimes employed under town centres which could not otherwise be undermined. Pillars of size approximately one-quarter of the depth of the working below the ground surface are left intact with panels equal to the pillar size extracted. Approximately one-half of the coal is worked by this method of mining and reductions in surface subsidence of up to 80% have been recorded.

Chapter 9

DISTRESS IN EXISTING STRUCTURES

As well as designing new buildings, architects and engineers are often called on for advice where existing buildings are showing signs of distress. Work with existing buildings presents more difficulty than work on new structures and dealing with buildings already suffering distress is particularly tricky. The first step is to diagnose the cause of the problem: only then can its seriousness and consequences be assessed and remedial measures suggested. The difficulty with diagnosis is that buildings move, crack and even collapse for a wide variety of reasons. This is an area where, more than most, experience is vital, and it must be recognised that in many cases it will not be possible to pinpoint the cause of a problem with absolute certainty. Correct diagnosis is often difficult to achieve where isolated small structures are affected because only a small amount of the funding will usually be available for the investigation.

The following sections give some reasons for movement and cracking of buildings, suggest how to detect the cause of movement and discuss the long term consequences of the various defects. This chapter should be considered to complement Chapter 7, although distress is not necessarily a consequence of problem soils.

STRUCTURAL DEFECTS

Signs of distress in a structure may arise entirely out of structural shortcomings and be completely unconnected with the foundations. Whether problems are basically structural or as a result of inadequate foundations is usually apparent from the nature of the problem. Where the cause is not obvious, close liaison between the structural and geotechnical engineers is essential, so that each is not blinkered by his own speciality. Discussion of defects due to purely structural causes is beyond the scope of this book.

MINING SUBSIDENCE

Most mining subsidence problems are due to coal extraction. This usually affects whole areas rather than individual buildings where modern longwall total extraction coal mining methods are used, and takes place within a specific interval of time. Problems may be more localised and sporadic, in the geographical and temporal senses, where traditional pillar-and-stall working was used. In this method, pillars of coal were left in the mined seams to support the roof. These pillars may fail within a short time after working, or after a century or more. The problem of coal mining areas is dealt with in Chapter 8.

Limestone mining can also produce localised and sporadic subsidence problems when pillar and stall working, similar to that used in coal mining, has been employed. This form of limestone extraction is adopted only when the outcrop is too narrow to allow normal quarrying or surface mining to take place. The pillars left in to support the roof are generally large enough to maintain stability of the workings but water ingress can lead rapidly to solution weathering and spalling of the limestone with unpredictable subsidence effects.

Another form of mining is solution mining. This method has been used in Cheshire to extract salt. Water is pumped into the ground to dissolve the salt and the resulting brine is then extracted from boreholes. What happens to the water after it enters the ground and where the salt is dissolved away cannot be determined with any precision, so pinpointing the affected areas is difficult. Nevertheless, many of the affected areas have been identified and documented; local records may be helpful in areas where this is known to be a problem.

One of the problems of diagnosis in mining areas is that distress to structures may arise from other causes instead of, or as well as, the mining activity. This adds complications to what may be an already complex situation.

BEARING CAPACITY FAILURE

Most of the problems that arise because of inadequate foundations are the result of excessive settlement rather than complete bearing capacity failure, which is rare. Foundations are at their least stable when the structure they support has just been completed so, if failure does not occur then, it is unlikely to occur later. This is discussed in Chapter 6 and more fully in Chapter 10. Where bearing capacity failures do occur in established structures they are almost invariably initiated by some outside change. Examples are a deep trench dug alongside a strip foundation, or water seeping beneath foundations and weakening the soil. Foundation failures sometimes occur in the downslope wall of old buildings on steep hillsides. The foundations, which may be nominal, are often constructed on soft soils which, with the added weight of the wall, possibly constructed of heavy stone masonry, are induced to creep downhill. The downslope wall tends to rotate away from the building and may eventually collapse.

Shear failure is usually accompanied by rotation of the foundation, although a similar effect can be produced by differential settlement. If necessary, soil samples can be taken from beside the foundation and triaxial and consolidation tests carried out to determine the cause of movement. Remedial measures depend very much on the individual case. Where softening of the foundation is a problem, it may be sufficient to install a drainage system to dry out the soil and arrest further movement. In some cases underpinning might be advisable or complete demolition of the affected wall may be necessary. One solution, often seen on old buildings, is to tie an affected wall to the rest of the building with steel tie rods. However, this alone does not tackle the cause of the problem.

CONSOLIDATION SETTLEMENTS

A common cause of cracking is high differential settlement due to consolidation of the ground beneath the foundations. There are several reasons why this might occur, each producing slightly different results and requiring different action.

Poor overall ground conditions

Detection. Poor overall ground conditions can be detected by standard site investigation methods, with particular emphasis on consolidation test results, so that settlement calculations can be made. With newer buildings, the original site investigation report should already contain sufficient information to draw a preliminary conclusion as to whether or not this is a likely cause of failure, although a supplementary investigation may be needed to confirm or reject other possibilities. If the original site investigation does show overall high compressibility to be a problem, the report should be checked to find out whether due allowance was made for this in the calculations and recommendations. The problem is likely to affect the whole building, with high total and differential settlements. If movement of the building is monitored, the rate of settlement should steadily decrease with time, although a long period of monitoring may be needed to show this. The situation is made worse where there is a wide variation in the foundation loading or founding depth within a single structure.

Long term consequences. Unless ground conditions are very poor, the soil should eventually consolidate and ground movements virtually cease. However, the resulting settlements may be unacceptably high and it may be several years before settlement is substantially complete: consolidation test results should provide answers to these questions. Where possible,it is probably better to monitor movement of the building before deciding on remedial work, to check that this is the cause and that the settlement estimates are correct; provided, of course, there is no immediate danger. With luck, remedial measures may be limited to repairing finishes and making up floor levels after the building has settled down.

Occurrence. Excessive consolidation settlements should not occur in modern buildings, whose foundation design should be appropriate for the ground conditions. This reduces the likelihood of this problem occurring in new buildings but, unfortunately, it is by no means unknown to find that consolidation settlement has not been properly allowed for in the foundation design, even where a site investigation has been carried out. Older buildings are often severely distorted as a result of consolidation settlement but, because the process takes place in the early years of a building's life, further movement from this cause should not be expected. If movement is continuing in an old building, it may be that the soil beneath the foundations is so overstressed that there is a danger of shear failure. Settlement or failure due to hillside creep can occur on sloping ground, as discussed earlier.

Isolated patches of poor ground

These could be either naturally occurring or brought about by previous development. For instance, a pond or damp depression may have silted up or been filled in, leaving an area containing weak infill material with softened clay beneath.

Detection. Distortion and cracking is likely to be limited to specific, well-defined areas. If a site investigation was carried out, the report may give some indication of a problem but it is possible that the areas of poor ground were missed. Boreholes or trial pits in the affected areas, and perhaps some in the unaffected areas as a check, with consolidation and strength testing of undisturbed samples, should help to determine the cause. A pocket penetrometer is particularly useful to pick out variable soil conditions in trial pits. This device is described in Chapter 3. Old maps and, particularly, old aerial photographs may be helpful to show earlier conditions on site. Reference to site records during construction of the affected building, and talks with site staff, if available, may be helpful.

Long term consequences. These are much the same as for consolidation settlement due to poor overall ground conditions.

Occurrence. It is most likely to be a problem in newer buildings: movements in older buildings will probably have virtually ceased after a few years if consolidation settlement is the sole cause.

Variations in subsoil conditions

The problems of building in variable ground conditions have been described in Chapter 7. Such variations may be natural, the result of dipping strata for instance, or man-made variations such as fill on part of the site.

Detection. Standard site investigation methods should reveal variations in ground conditions. Previous site investigation reports should be checked. A geological map of the area will often prove helpful. Distress is usually limited mainly to a particular area, which is likely to be in the neighbourhood of a junction or transition zone between different ground conditions.

Long term consequences. So much depends on the circumstances of each individual site that it is not possible to generalise.

Occurrence. Problems usually arise soon after construction so distress from this cause usually leads to difficulties with newer buildings.

Water under the foundations

Water from a broken pipe or similar source can weaken the foundation soil and cause consolidation settlement.

Detection. Only limited areas are likely to be affected. If water is suspected to be the cause of movement, water pipes and drains should be checked, as far as possible, for leaks. Any other source of water, such as surface runoff, should also be considered, with special attention to any changes that may have taken place shortly before the problem became apparent. Soil samples taken from trial pits should be tested for moisture content to check whether there are any significant variations.

Long term consequences. Once the source of water has been detected and stopped, further problems are unlikely although it may take some time for movement to effectively cease.

Occurrence. This can occur at any time, in new or old buildings. However, where water seepage does lead to movement, this may also be the result of swelling of expansive clays, discussed later in this chapter and in Chapter 7, rather than of consolidation settlements.

Bad practice during construction

Problems of this type usually arise because foundation excavations were left open for an extended period and were allowed to become flooded. The clay beneath swells, leading to excessive and uneven settlement of the structure early in its life. Initial shrinkage of a clay left exposed in hot weather, with subsequent swelling, is also a possibility, but less likely to be a problem.

Detection. Once the building has been constructed and the soil beneath the foundation has had time to return to its original moisture content, it is difficult or impossible to tell whether bad construction practice is the cause of distress except by inference. The likelihood is increased if foundations are constructed in wet weather, especially in the winter months, and if site supervision or the contractor's standards are poor. Site records and discussions with site staff may be useful.

Long term consequences. The depth of ground affected will usually not be great so that, with luck, most of the movement will be over within a year or two, and finishes can be repaired, but it is not possible to generalise.

Occurrence. The problem occurs in new buildings, especially where there is poor supervision. In this respect, private housebuilding, where there is usually no client supervision, must give cause for concern.

SWELLING AND SHRINKAGE OF EXPANSIVE CLAYS

Severe ground movements can occur where foundations on expansive clays are not taken deep enough. This is discussed in Chapter 7.

Detection. In Britain, cracking usually occurs after a prolonged dry spell. One side of the building may be affected more than another. Other buildings in the area may be similarly affected. Trees planted close to the building can have a particularly adverse effect. Liquid and plastic limit tests should be carried out on samples of the soil to check whether it is likely to be expansive. If necessary, linear shrinkage or other specialised tests may be requested but this is not usually necessary. Boreholes, trial pits or hand auger holes alongside the building will enable moisture content samples to be taken at frequent intervals of depth, from several locations. A plot of moisture content against depth will show whether there has been significant drying out of the upper soil and, if so, to what depth.

Long term consequences. If nothing is done, it is likely that, with the advent of wetter weather, movement will be partly reversed and cracks will become narrower. There is then not likely to be further substantial movement until the next dry spell. Low cost protection includes the removal of trees and the provision of an impermeable apron to protect the building from seasonal moisture content variations, as discussed in Chapter 7. Underpinning has been frequently used to prevent further problems.

Occurrence. The problem can arise wherever shallow foundations are used on expansive soils. London clay is a notorious example. This affliction can affect new or old buildings.

GENERAL

It can be seen that there are many possible causes of distress to buildings, even when only geotechnical aspects are considered. Quite different underlying causes can give rise to similar effects so that it is difficult even for a specialist to detect the cause with any certainty. Yet this may be vital if appropriate remedial measures are to be taken. Site investigation work is aimed at checking soil strength and consolidation properties, variations of soils or soil properties from one part of the site to another, moisture content and its variation with depth and location, and assessing the swelling potential. Results must be carefully correlated and compared with the observed pattern of distress. In this respect, it can be helpful if movement of the building is monitored over a period of time. There is more chance of coming to the correct conclusion if time is available for this.

Chapter 10

PRINCIPLES OF SOIL MECHANICS

IDEALISED SOIL STRUCTURE

Soil is a complex material which contains solid, liquid and gaseous phases. The solid material forms the soil skeleton and is made up of small rock fragments which have undergone varying degrees of chemical alteration. The voids between the solid particles are filled with water and air, as illustrated in Figure 10.1 (a). The

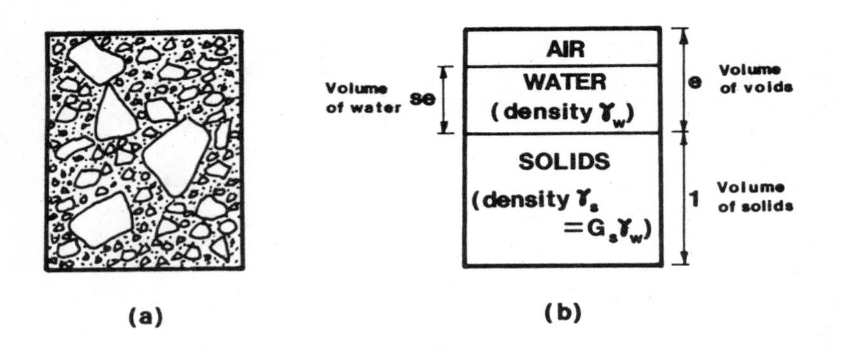

Figure 10.1 The model soil sample: (a) actual soil sample; (b) idealised model.

properties of the soil are affected not only by the materials that make up the solids and their shapes and grading, but also by the proportion of the soil that is occupied by the voids and by the proportions of air and water in those voids. To simplify ideas for calculation it is convenient to imagine that the air, water and solid constituents have been separated out neatly as shown in Figure 10.1 (b). To simplify matters further, it is usually assumed that a sample of soil has been chosen that contains a unit volume of solid material, as illustrated. This simplified model of soil is known as the model soil sample. The proportion of voids in a soil can be defined in two ways:

$$\text{voids ratio, } e = \frac{\text{volume of voids}}{\text{volume of solids}}$$

$$\text{porosity, } n = \frac{\text{volume of voids}}{\text{total volume}}$$

In soil mechanics it is more usual to work with the voids ratio. It can be seen that the model soil sample, with unit volume of solids, contains a volume, e, of voids. Since its total volume will be (1 + e) then, from the definition for porosity,

$$n = \frac{e}{1 + e}$$

In general, some proportion, s, of the voids will be filled with water. This is known as the degree of saturation of the soil. Thus, we can define

$$\text{degree of saturation, } s = \frac{\text{volume of water}}{\text{volume of voids}} .$$

In the model soil sample the volume of water will therefore be s.e.

Although the voids ratio and degree of saturation are useful concepts when considering the volumetric proportions of solid, water and air which make up the soil, they cannot be measured directly in the laboratory, where it is much more convenient to measure weights or masses. We can define moisture (or water) content in terms of weighings as:

$$\text{moisture content, } m = \frac{\text{mass of water}}{\text{mass of solids}} .$$

If the density of water is γ_w and the density of solids is γ_s (= $G_s.\gamma_w$, where G_s is the specific gravity of the solids) then, since the volume of water is s.e and the volume of solids is 1:

$$m = \frac{\text{volume of water.}\ \gamma_w}{\text{volume of solids.}G_s.\gamma_w} = \frac{s.e}{G_s}$$

or

$$s.e = m.G_s$$

The density of the soil, γ is defined as

$$\gamma = \frac{\text{mass of solids + mass of water}}{\text{total volume}}$$

The mass of water can be expressed in two ways: either as the volume of water multiplied by its density, $s.e.\gamma_w$, or as a proportion of the mass of solids using the definition of moisture content, $m.G_s.\gamma_w$. Thus

$$\gamma = \frac{(G_s + s.e)\gamma_w}{1 + e} = \frac{G_s(1 + m)\gamma_w}{1 + e}$$

Density defined in this way is known as the "bulk density". For a saturated soil, s = 1 and the density becomes the "saturated density". The density will obviously change with moisture content even though the soil skeleton itself remains fixed.

As a measure of the state of packing of the soil, the "dry density" is used. This ignores the contribution of the water contained in the soil and is defined as

$$\gamma_d = \frac{\text{mass of solids}}{\text{total volume}} = \frac{G_s.\gamma_w}{1 + e}$$

Comparing bulk and dry densities:

$$\frac{\gamma_d}{\gamma} = \frac{G_s.\gamma_w}{1 + e} \cdot \frac{1 + e}{G_s(1 + m)\gamma_w} = \frac{1}{1 + m}$$

giving

$$\gamma_d = \frac{\gamma}{1 + m}$$

Thus, if the bulk density and moisture content of a soil are measured in the laboratory, the dry density can be calculated. If the specific gravity of the soil solids is measured then the voids ratio and degree of saturation can be calculated.

A submerged soil will experience an upthrust due to the buoyancy effect in water and its effective density will be the "submerged density", γ_{sub}, where

$$\gamma_{sub} = \gamma - \gamma_w$$

SHEAR STRENGTH

When a material is subjected to progressively higher stresses it will eventually break, or fail. Stresses may be compressive, tensile, or shearing, as illustrated in Figure 10.2. Purely compressive

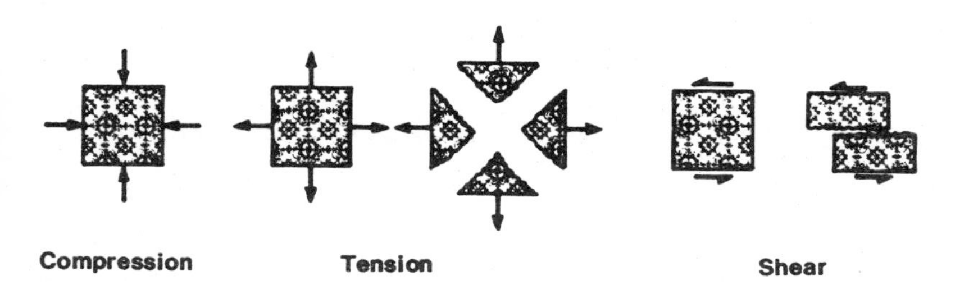

Figure 10.2 Stresses acting on a solid and the resultant modes of failure.

stresses do not lead to failure, since by their nature they simply squeeze the material inwards on itself. Tensile stresses can lead to tensile failure, as illustrated, but their occurrence is limited and designs usually assume that a soil has no tensile strength. Shear stresses can lead to shear failure and this is almost always the mode of failure of soil masses.

Shear stresses are caused in soils by variations in compressive stress with direction, as indicated in Figure 10.3, which represents stresses on a cube of material within a soil mass. In general, there will be both normal and shear stresses acting on the sides of the cube but by orientating it so that the sides face certain directions the shear stresses will become zero and the normal stresses will reach minimum and maximum values, σ_1 and σ_3 respectively. These are the

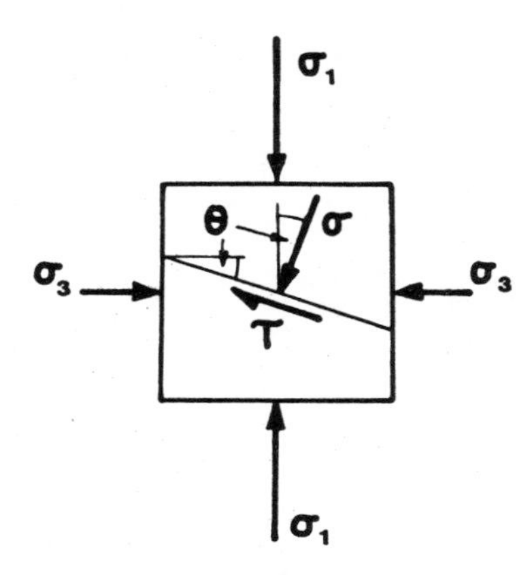

Figure 10.3 Shear stresses, which result from variations in compressive stress with direction.

principal stresses. There is also an intermediate principal stress, σ_2, acting on the cube normal to the plane of the paper. In the ground the maximum principal stress usually acts vertically or near vertically and is equal to the overburden pressure plus any applied surface loading. Stresses σ_2 and σ_3 are usually assumed equal and represent the lateral thrust of the ground. The difference in the magnitudes of the principal stresses leads to shear stresses within the cube. On any plane within the cube there will generally be both normal and shear stresses, as illustrated. The relative proportions of these depend on the direction chosen; on the value of θ. In the triaxial test, σ_1 is steadily increased while σ_3 is kept constant and eventually the soil shears along a plane at some angle θ.

A convenient method of representing stresses within a material is to plot shear stress against direct stress, as shown in Figure 10.4. The principal stresses lie on the axis. The direct and shear stresses on any plane at an angle θ to the maximum principal stress plane are obtained by the construction shown. This is known as the Mohr circle construction. It can be seen that the maximum shear stress occurs where $2\theta = 90^\circ$; at $\theta = 45^\circ$ to the σ_1 plane, and that its value is equal to the radius of the Mohr circle,

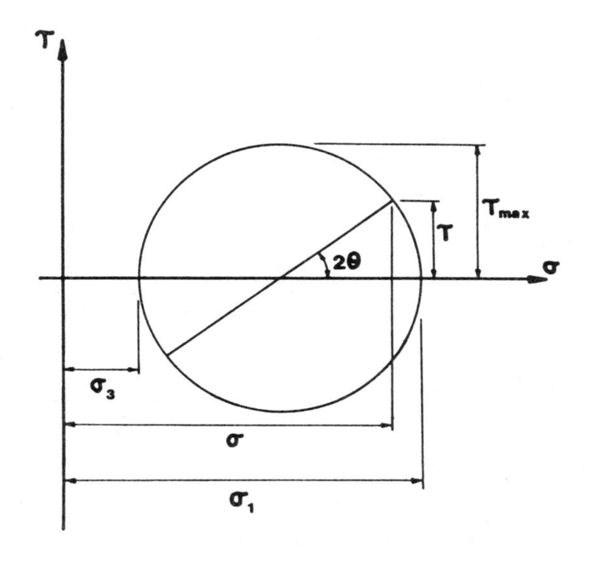

Figure 10.4 Mohr circle construction.

$\frac{1}{2}(\sigma 1 - \sigma 3)$.

The Mohr diagram can also be used to represent the stress combinations that would lead to failure. In a purely frictional material, where the shear stress at failure along any plane is a proportion of the normal stress on that plane, the failure condition is represented by a line through the origin, as shown in Figure 10.5. Circle I on the diagram represents a safe state of stress, since all combinations of stress around the circle are within the allowable limits. Circle II represents a state of stress at failure. No circle can cross the failure line because the soil would have already yielded before this stress condition was reached. The slope of the line represents the coefficient of friction within the material - the ratio of shear stress to normal stress at failure. It is usually expressed as angle ϕ, the angle of shearing resistance. This is analogous to the angle of friction between two surfaces. Following this analogy, failure will occur along any plane when the shear stress reaches a value

$$\tau_{max} = \sigma \tan \phi$$

where σ is the normal (direct) stress on that plane.

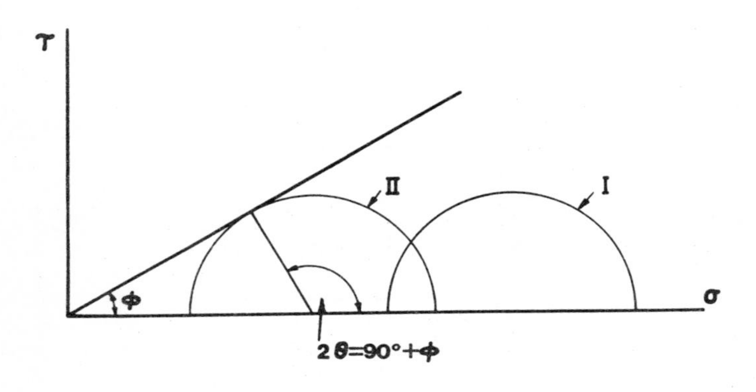

Figure 10.5 Failure envelope for a purely frictional material, plotted on the Mohr diagram.

Note that in circle II the critical plane along which failure will take place is not along the direction of maximum shear but at an angle $(45^{0}+\frac{1}{2}\phi)$ (ie $\frac{1}{2}(90^{0}+\phi)$) to the plane on which σ_1 acts. For a purely cohesive material, in which $\phi = 0$ and the shear strength is independent of the values of σ_1 and σ_3, the failure condition would plot as a horizontal line as shown in Figure 10.6. In this case, the material has a fixed shear strength, c, and failure will occur on a plane when the maximum shear stress reaches the value

$$\tau_{max} = c$$

regardless of the normal stress on the plane. Note that, for a purely cohesive material the failure plane should correspond with the plane of maximum shear, at 45^{0} to the principal stress directions. In practice, this is not true for soils because of the effects of pore water pressures. To produce a general model for

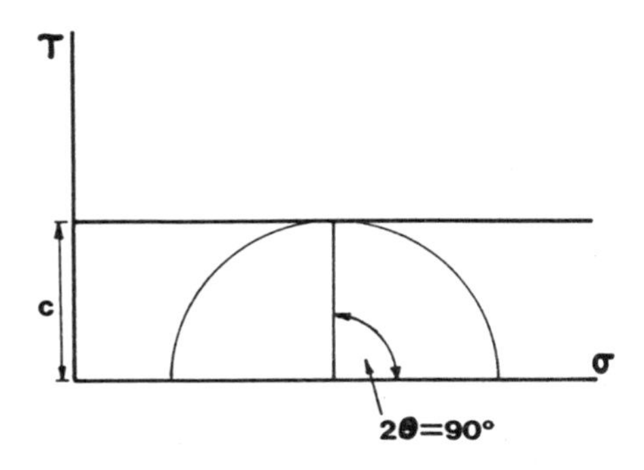

Figure 10.6 Failure envelope for a purely cohesive material, plotted on the Mohr diagram.

the failure of soils we can consider a soil whose strength is made up of both friction and cohesion components. The maximum shear stress on any plane will then be

$$\tau_{max} = c + \sigma \tan \phi$$

This is known as the Mohr-Coulomb failure criterion and is the most commonly used failure theory in soil mechanics. The Mohr-Coulomb failure line is represented on the Mohr diagram as shown in Figure 10.7. The critical plane, along which failure will occur is at $(45^o+\frac{1}{2}\phi)$ to the plane on which σ_1 acts, as it is for a purely frictional material. Mohr circle plots give a useful tool in helping to understand the Mohr-Coulomb failure theory and are used with triaxial testing to obtain c and ϕ values.

TOTAL AND EFFECTIVE STRESSES

Because soil is not the simple homogeneous material which is assumed when calculating stresses by classical elastic theory, stress conditions are rather more complicated than they at first appear. Stress is actually transmitted through the soil skeleton as a

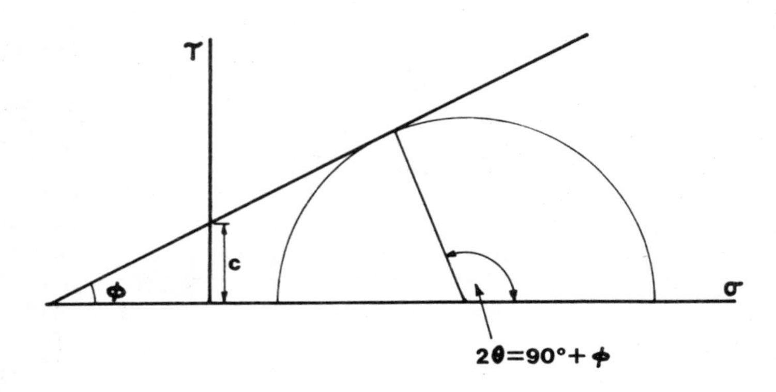

Figure 10.7 Failure envelope for a material possessing both internal friction and cohesion: the Mohr-Coulomb failure condition.

multitude of small forces acting at the points of contact between soil particles. This makes the response of the soil to stresses quite different from the response of the materials which make up the soil grains. Compressive stresses distort the soil grains elastically. This distortion is immediate and recoverable but is usually too small to be of concern. In clay soils, compressive stress can also slowly drive out water which is loosely bonded to the clay minerals by electrostatic forces. Layers of such water surrounding and within the clay mineral molecules are known as the adsorption complex. The process takes time, so gives the clay time-dependent properties but the distortion is usually recoverable: water will be re-adsorbed and the clay minerals will swell once the stress is removed. The major part of the displacement caused by applied stresses results from the grains moving closer together, with a consequent reduction in the volume of voids. Once such movement has occurred it will not be reversed if stresses are removed.

The rate at which this packing together of the soil particles occurs depends on the ground water conditions. In a dry or partially saturated soil, the air in the voids is easily compressed and quickly expelled so the process is almost immediate. In a saturated or nearly saturated soil the water filling the voids is relatively incompressible and a rearrangement of the soil particles, with a reduction in voids, cannot take place until water has had time to flow out of the soil The rate at which this happens depends on the permeability of the soil and the distance the pore water has to travel. This is the mechanism of consolidation which causes the slow settlement of structures, since most soils are saturated except for a relatively thin surface layer.

Because of the time lag required for saturated soils to compress, the soil skeleton cannot immediately take up an applied compressive stress. Instead, the extra compressive stress is taken by the pore water. Thus, the compressive stress, or pressure, within the soil is made up of two components; that carried by the soil skeleton and that due to the pressure of the pore water. The loading experienced by the soil skeleton is known as the "effective stress" and that carried by the soil mass as a whole, including the pore pressure, is known as the "total stress". As water seeps out of a soil in response to an applied pressure, so the effective stress on the soil skeleton will slowly increase until eventually all the extra applied pressure is carried as effective stress.

The way in which stresses are distributed in a soil can be illustrated by considering the situation shown in Figure 10.8. In zone A the soil may be dry or only partially saturated and all stresses are carried by the soil skeleton. In zone B the soil is saturated but the water is not under any pressure: it can be thought of as resting on the soil particles themselves, sucked up by capillary action. As with zone A, all stresses will be carried by the soil skeleton. In zone C, below the standing water table, the water is under normal hydrostatic pressure, as if it existed in a tank. For simplicity, let us assume that the soil all has the same bulk density γ. Consider two pressure transducers buried at some depth z in zone C. One is embedded directly in the soil and measures the total vertical pressure $\sigma_v = \gamma . z$ due to the weight of soil above it. The second transducer is surrounded by a porous pot to prevent the soil pressing against it, so that it acts as a piezometer. This will record only the pore water pressure, u, where, in this case $u = h . \gamma_w$. The

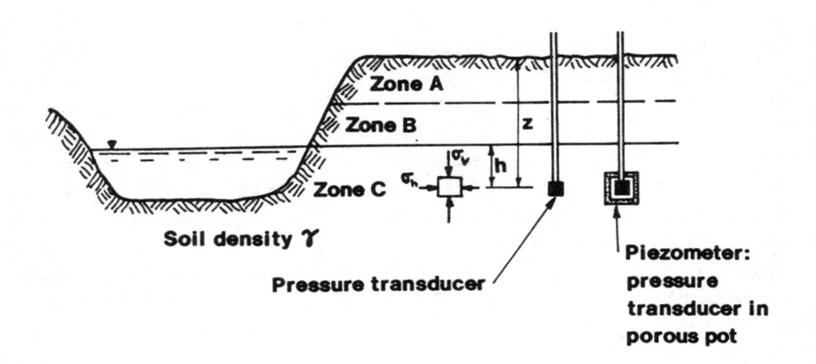

Figure 10.8 Total and effective stresses within a soil mass.

difference between these two values represents the pressures carried by the soil skeleton; the effective vertical pressure σ'_v. Thus

$$\sigma'_v = \sigma_v - u$$

In general, for direct stress in any direction

$$\sigma' = \sigma - u$$

i.e.
effective stress = total stress - pore water pressure

In zones A and B, there is no pore pressure so the effective (soil skeleton) pressures will be equal to the total pressures.

As described earlier, if an extra pressure is applied to the soil, by a new building for instance, the resulting extra pressure in the saturated zones B and C will be carried initially by the pore water but will be slowly transferred to the soil skeleton as the water seeps away.

So far, we have considered only direct stresses but the soil will also experience shear stress. Small shear stresses will be present in the soil's natural condition due to the fact that horizontal pressures are usually smaller than vertical pressures. More importantly, shear stresses will occur beneath building foundations due to their loading. In contrast to direct stresses, shear stresses and shear movements can develop in the soil skeleton without appreciable change in volume, so there is no time delay while water seeps into or out of the voids and shear stresses are carried immediately by the soil skeleton. The pore water cannot carry shear stresses since it has no shear strength. Although loading from a foundation will produce an increase in both normal and shear total stresses, when considered in terms of effective stresses, only the shear stresses will increase. Since shear strength is produced by resistance between soil grains, which is primarily frictional so depends on normal stresses, there will be no immediate change in the shear strength of the soil. Only later, as pore water pressures dissipate and consolidation takes place, will shear strength increase. This means that the most critical time for foundations occurs at the end-of-construction period and it is for this reason that quick undrained shear strength parameters are used in foundation design. The analysis could be carried out using effective stress parameters, obtained from consolidated undrained tests, with an allowance for the pore pressures developed beneath the foundation; but the use of undrained strength values effectively takes care of the pore water pressures without the need to calculate them or to measure them in laboratory tests.

The situation is quite different in excavations. In this case, the removal of soil allows the sides of the excavation to swell. The pore water now prevents immediate swelling, sucking the soil particles together until water has time to flow in and allow expansion of the voids. Thus, the peak shear strength occurs initially, when pore pressures are at their lowest, and slowly decreases with time. This is why excavations are often stable initially but collapse after some time, and why they can often be cut with vertical sides provided they are not left open too long. Short term stability can be calculated using quick undrained shear strength parameters, but for the more critical long term stability an effective stress analysis must be used, based on drained tests, with an allowance for the anticipated long term pore water pressures. The long term stability of natural slopes and embankments is carried out in the same way. However, in the case of embankments, initial stability may be more critical because of pore pressures developed by the added weight of material as the embankment is built up and by the compaction process. For large embankments it is usual to monitor pore pressures during construction and if necessary to reduce the rate of construction to limit pore pressures to safe levels. With retaining walls it is usually the long term stability which is critical but, depending on the type of backfill and the method of construction used, it may be necessary to check both short and long term stability.

Returning now to consideration of the immediate response of a saturated soil to stresses, we can see that, even for an intrinsically frictional material, since external loading initially produces no change in normal stresses in the soil skeleton, there will be no change in shear strength. In the triaxial test, for instance, the shear strength of a saturated clay is independent of cell pressure because any change in cell pressure is immediately balanced out by an equal change in pore pressure within the soil sample, leaving normal stresses within the soil skeleton unchanged. This leads to the well-known phenomenon of a "purely cohesive" soil which has a zero angle of shearing resistance. The cohesion is not really a property of the soil skeleton but a consequence of the water pressures developed in the voids. Given time, and allowed to drain, even a so-called purely cohesive clay would behave as a frictional material, with little or no cohesion.

CONSOLIDATION SETTLEMENT

As described earlier in this chapter, consolidation settlement occurs primarily as a result of a rearrangement of the soil grains in response to an applied load. This process can be illustrated in terms of the model soil sample, as shown in Figure 10.9. The consolidation characteristics of a soil are defined in terms of the coefficient of volume compressibility, m_v, which is a measure of the amount of consolidation; and the coefficient of consolidation, c_v, which is a measure of the rate of consolidation.

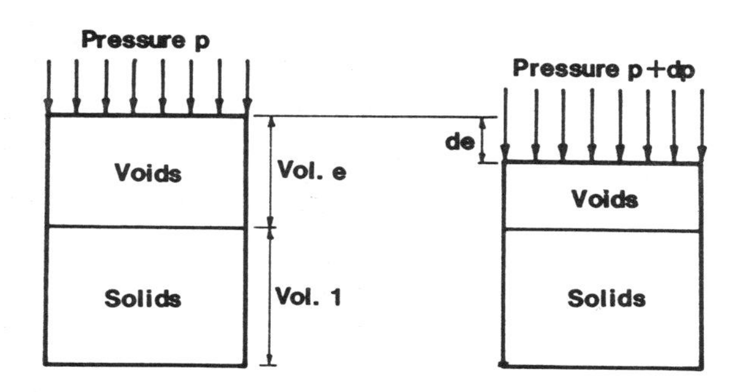

Figure 10.9 The consolidation process illustrated by use of the model soil sample.

The coefficient of volume compressibility is defined as the decrease in unit volume per unit increase in pressure. From the model soil sample illustrated, it is given by

$$m_v = \frac{\text{volume change}}{\text{original volume}} \div \text{pressure change}$$

Thus

$$m_v = \frac{de}{1 + e}/dp = \frac{1}{1 + e} \cdot \frac{de}{dp}$$

In the consolidation test the soil is constrained laterally so that the cross-sectional area remains constant. Thus, the volume of the specimen is proportional to its thickness, and the coefficient of compressibility can be expressed as

$$m_v = \frac{\text{change in thickness}}{\text{original thickness}} \div \text{pressure change}$$

or

$$m_v = \frac{dh}{h}/dp = \frac{1}{h} \cdot \frac{dh}{dp}$$

From consolidation test results the specimen thickness

is usually plotted (to a linear scale) against the effective pressure (to a logarithmic scale), as illustrated in Figure 10.10. As an alternative, some people prefer to calculate the voids ratio, e, of the specimen and plot e against pressure. As discussed above, the two methods are exactly equivalent, since height is proportional to volume. The e/p plot expresses the relationship in a more fundamental form but the direct plot of thickness against pressure avoids unnecessary calculations and the need to measure or assume a value for G_s, the specific gravity of the soil solids, which is needed to calculate the voids ratio. Values of m_v can be obtained from the plot.

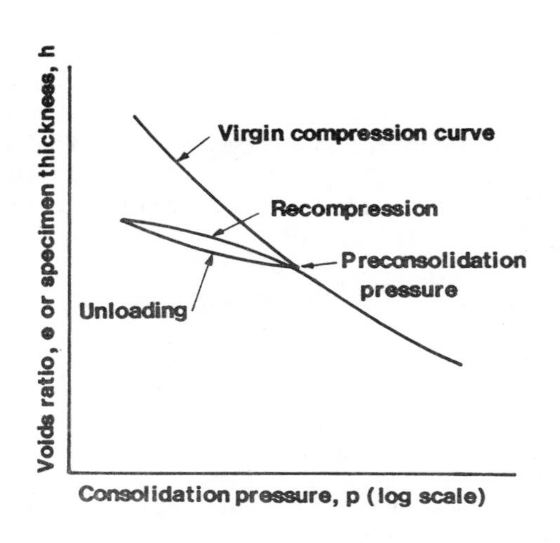

Figure 10.10 Typical consolidation test results.

The value obtained will depend on the pressure range chosen and it is important to choose the pressure range that corresponds to field conditions. Beneath a proposed foundation the initial pressure in the soil at a given depth will be the overburden pressure and the final pressure will be the overburden pressure plus the extra pressure due to the foundation loading: this is the pressure range used when calculating m_v values.

The curve shown in Figure 10.10 includes an unloading stage and a recompression. As might be expected, once a soil has been loaded and then unloaded, subsequent recompression produces less consolidation than was experienced during the initial loading. Once the preconsolidation load is exceeded, consolidation follows the virgin curve again. This is an important consideration when dealing with overconsolidated clays, which have experienced higher consolidation pressures during their geological history than occur in present time. This may have been caused by depths of overburden that have been removed by weathering; by a covering of ice during an ice age; or by a lowering of the water table and partial drying out at some stage. Such soils will be relatively stiff at loadings below the preconsolidation pressure but once this is exceeded the m_v value may increase. Failure to take this into account can lead to underestimation or overestimation of settlement.

Once a load increment is applied to the soil in the consolidation test, values of specimen thickness are taken at set times so that a plot can be made of thickness against time. The time scale is plotted as the square root of time or as the logarithm of time. From these plots, values of c_v can be found. The variation of c_v values with load increment is much less marked than it is for m_v values. The time for a required degree of settlement (the amount of settlement as a proportion of the final settlement) to take place in a layer of clay can be calculated from the expression

$$t = \frac{T_v \, d^2}{c_v}$$

where d is the maximum length of the drainage path (equal to the layer thickness if drainage is from the top or bottom only but equal to half the layer thickness if drainage is from both top and bottom).
and T_v is the basic time factor.

Values of T_v depend on the degree of consolidation, the drainage conditions and the pressure distribution, as indicated in Table 10.1. Settlement is usually

TABLE 10.1
VALUES OF TIME FACTOR T_v

U	T_v			Drainage conditions and pressure distributions		
	case 1	case 2	case 3	case 1*	case 2	case 3
0.1	0.008	0.047	0.003			
0.2	0.031	0.100	0.009			
0.3	0.071	0.158	0.024			
0.4	0.126	0.221	0.048			
0.5	0.197	0.294	0.092			
0.6	0.287	0.383	0.160			
0.7	0.403	0.500	0.271	Any pressure distribution, drainage top and bottom	Decreasing pressure, drainage at bottom only	Decreasing pressure, drainage at top only
0.8	0.567	0.665	0.440			
0.9	0.848	0.940	0.720			

*Case 1 may be used for uniform pressure distribution with drainage at top or bottom only.

considered to be substantially complete when the degree of consolidation is 90% (U = 0.9 in the table). To achieve 100% of final settlement would theoretically take an infinite time.

Some soils, such as fissured clays and stratified estuarine silts have a macrostructure which allows water to drain away relatively freely. Because of this, consolidation takes place at a much higher rate than is predicted by c_v values obtained in the laboratory test: time factors of 10:1 are common and factors as high as 100:1 are not unknown. In such soils it is better to calculate c_v from the results of field permeability tests using the formula

$$c_v = \frac{k}{m_v \cdot \gamma_w}$$

where k is the field coefficient of permeability
γ_w is the density of water
m_v is the coefficient of volume compressibility, obtained from laboratory tests.

Consolidation theory applies only to saturated soils. For partially saturated soils m_v values may be obtained but c_v values would be meaningless. This is not usually a problem because settlements are rapid in partially saturated soils. To simulate field conditions and to assess the effects of saturation on such soils, specially modified consolidation tests may be needed.

Chapter 11

FOUNDATION DESIGN

ULTIMATE AND ALLOWABLE BEARING CAPACITIES

Foundations can fail in two ways: the foundation can punch into the ground, shearing a block of soil beneath it as shown in Figure 11.1(a) and (b); or it can settle excessively, causing distress to the structure it supports without total shear failure occurring, as shown in Figure 11.1(c). As discussed in Chapters 1 and 4, both of these factors must be taken into account when selecting an allowable bearing pressure.

The first step in deciding on a suitable allowable bearing pressure is to calculate the ultimate bearing capacity. This represents the point at which total collapse would occur. In practice, partial shear failure occurs around the edges of a foundation at a lower load than that required to cause total collapse. To avoid this, and for the reasons discussed in Chapter 1, the ultimate bearing capacity is divided by a relatively high factor of safety, of 2.5 or 3, to obtain an initial estimate of the allowable bearing pressure. Preliminary foundation widths can then be decided on. The second step is to calculate the settlement which would result from the preliminary design. If this is acceptably small then the allowable bearing pressure value is accepted. If predicted settlement is too large then the allowable bearing pressure is reduced to limit settlement to acceptable values. In practice, with a fixed weight of structure to support, this may not be all that simple, for reasons discussed later.

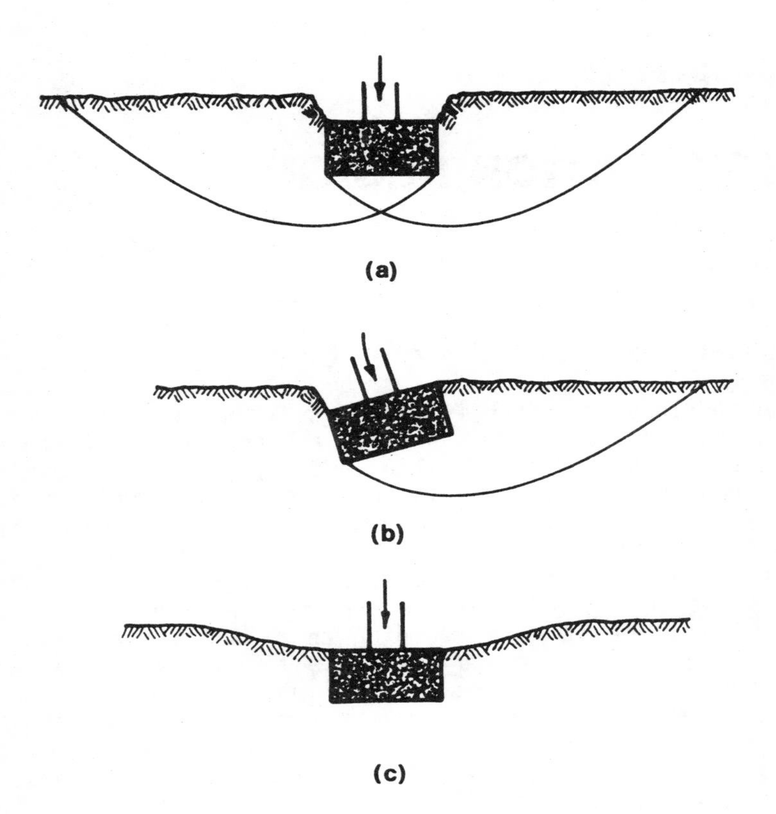

Figure 11.1 Failure modes of foundations: (a) symmetric shear failure (rare); (b) asymmetric shear failure; (c) excessive settlement.

METHODS OF ESTIMATING ULTIMATE BEARING CAPACITY

All the methods in common use assume that the soil beneath the foundation shears along a surface of a particular shape, as illustrated in Figure 11.1. In theory, failure by either of the mechanisms shown in (a) or (b) is equally likely, since the calculated failure loads are the same, but in practice failure mechanism (b) is more likely because of variability within the soil. As described in Chapter 10, the most critical time for the stability of foundations on clays is at the end of construction, before pore pressures beneath the foundation have had time to dissipate. For this reason, foundations on clays are designed using quick undrained triaxial or shear box tests. Drainage from granular soils is rapid so the distinction between the soil's immediate and long term responses vanishes and drained values are appropriate. Strength parameters for granular soils may be obtained from shear box tests but are more usually inferred from field tests such as the standard penetration test. Figure 11.2 shows a correlation between relative density, N-values obtained from the standard penetration test, and the angle of shearing resistance, ϕ, for sands.

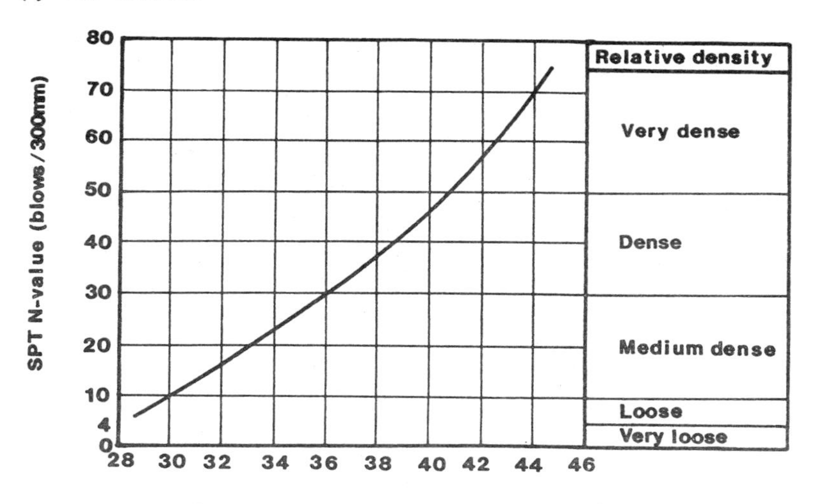

Figure 11.2 Correlation of values of ϕ with SPT N-values obtained by Peck, Hanson and Thornburn.

A number of methods exist for calculating ultimate bearing capacity but the most commonly used are those proposed by Terzaghi, Meyerhof and Hansen.

Terzaghi's method

From the basic premise of shear failure along an

assumed surface, Terzaghi developed the following expression for the net ultimate bearing capacity, q_{nu}, of a shallow strip foundation

$$q_{nu} = cN_c + p_o(N_q - 1) + \tfrac{1}{2}.\gamma.BN_\gamma$$

where γ is the bulk density of the soil below the foundation level(1)
c is the cohesive strength of the soil
p_o is the effective overburden pressure at foundation level(2)
B is the foundation width
N_c, N_q and N_γ are bearing capacity factors, obtained from Figure 11.3(3).

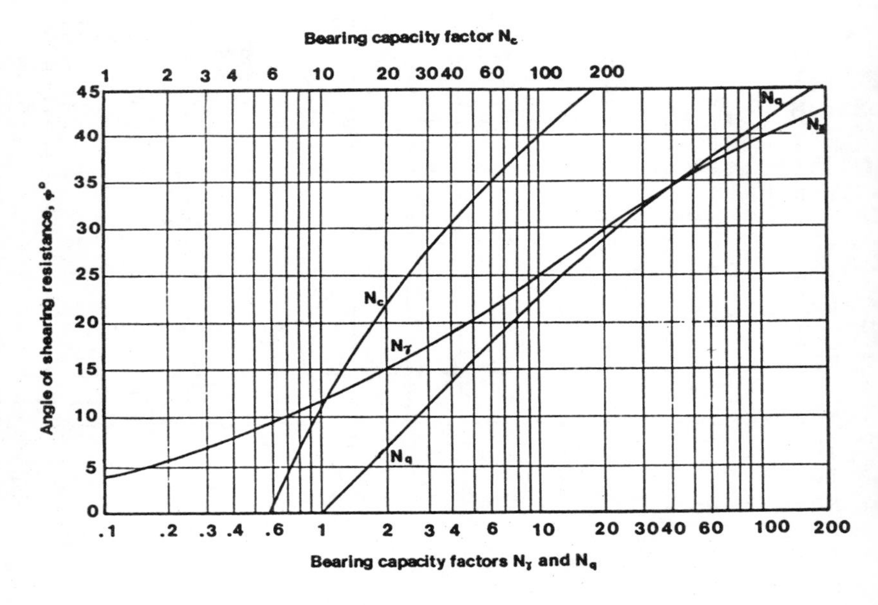

Figure 11.3 Terzaghi's bearing capacity factors.

Notes:

(1) If the water table is at or above founding level then the value of γ used in the bearing capacity calculations is the submerged density.

(2) The effective overburden pressure is normally γz, where z is the depth of the foundation, but if the

water table is above the base of the foundation then the submerged density is used for the soil below the water table.

(3) Before the bearing capacity coefficients can be obtained from Figure 11.3, a value for the angle of shearing resistance must be determined. As described above for sands, it is usually estimated from Figure 11.2.

For a square or circular foundation the equation is modified to:

$$q_{nu} = 1.3\ cN_c + p_o\ (N_q - 1) + 0.4\ \gamma\ BN_\gamma$$

Terzaghi's method ignores the shear strength of soil above the base of the foundation so bearing capacity is underestimated. The error is small for shallow foundations on granular soil and, since the method is easy to use, it is the most popular method of calculating bearing capacity in these conditions. It can also be used to calculate the bearing capacity of shallow foundations on clay soils but the error is larger, leading to conservative designs, and it is more usual to use Meyerhof's methods for foundations on clays. It is not suitable where deep foundations are to be used (where the founding depth is greater than the foundation width) because it leads to excessively conservative designs.

Meyerhof's method

Meyerhof's approach is similar to Terzaghi's but takes into account the shear strength of the soil above the founding depth and the different modes of failure of shallow and deep foundations. The bearing capacity factors also take into account the shape of the foundation. This makes the method more accurate and more comprehensive than Terzaghi's method but less simple to use. The basic equation for bearing capacity is:

$$q_{nu} = cN_c + p_o\ (N_q - 1) + \tfrac{1}{2}\ \gamma\ BN_\gamma$$

where N_c, N_q and N_γ are Meyerhof's bearing capacity factors.

Although this equation has the same form as Terzaghi's equation for a strip foundation the values of the bearing capacity factors used in it depend on the shape and depth of the foundation. Values of N_c, N_q and N_γ for a strip foundation are given in Figure 11.4. For rectangular, square or circular foundations these values must be multiplied by a shape factor λ. Values of λ are given in Figure 11.5. If the water table is at or above the founding level, bulk density γ is replaced in the equation by the submerged density. The value of p_o is obtained as described for Terzaghi's equations.

For saturated clays, using a total stress analysis, the soil typically behaves as a purely cohesive material. In this case, the bearing capacity equation reduces to:

$$q_{nu} = cN_c$$

where bearing capacity factor N_c is obtained from Figure 11.6. It is usual to use the curves obtained from experimental results rather than those based on theoretical analysis. Meyerhof's method is usually preferred for foundations on clays since it is easy to use and models the conditions more accurately than Terzaghi's method. It is also the most common method for deep foundations.

Hansen's method

This is much more comprehensive than Meyerhof's method, although the basic approach is similar. The method is little used in Britain and North America, where Terzaghi's and Meyerhof's methods are preferred but it is popular in Western Europe. Its great advantage is that it takes into account many more factors than the other two methods described so can be used to solve a wider range of foundation problems. Hansen's method results in a more conservative design than Meyerhof's method. The general equation is:

$$q_{nu} = cN_c s_c d_c i_c b_c g_c + p_o N_q s_q d_q i_q b_q g_q + \tfrac{1}{2}\gamma B N_\gamma s_\gamma d_\gamma i_\gamma b_\gamma g_\gamma - p$$

where c is the cohesion of the soil

p_o is the effective overburden pressure at the

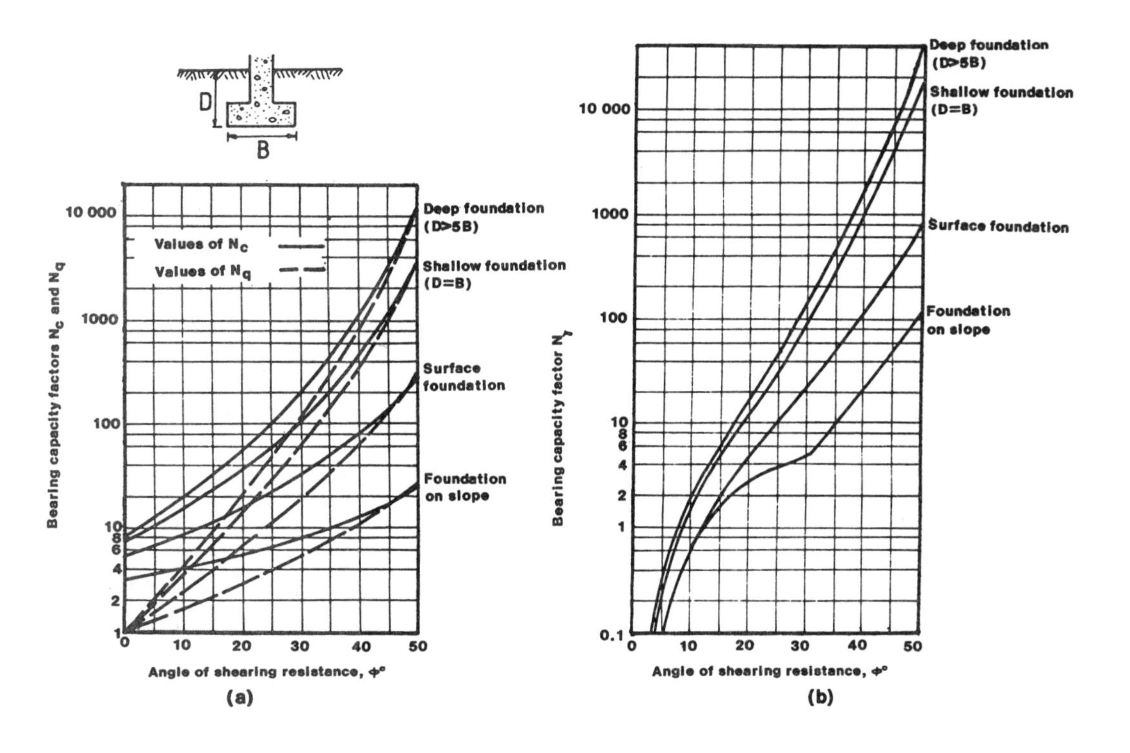

Figure 11.4 Meyerhof's bearing capacity factors for a strip foundation.

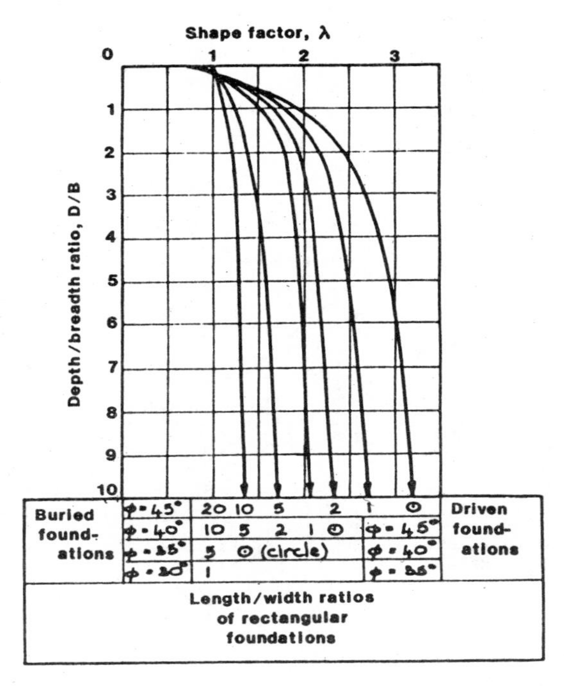

Figure 11.5 Values of factor λ for rectangular and circular foundations.

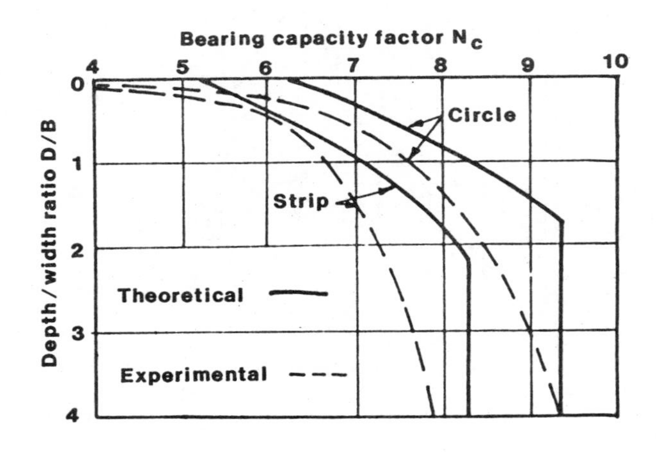

Figure 11.6 Values of N_c for foundations on a purely cohesive soil.

founding level

p is the total overburden pressure at founding level

γ is the bulk density of soil below the foundation

N_c, N_q and N_γ are bearing capacity factors
s_c, s_q and s_γ are shape factors
d_c, d_q and d_γ are depth factors
i_c, i_q and i_γ are load inclination factors
b_c, b_q and b_γ are base inclination factors
g_c, g_q and g_γ are ground surface inclination factors

Evaluation of some of these factors is quite complicated, making the method cumbersome to use for the majority of foundation problems which have simple geometries and carry vertical loads.

ECCENTRIC LOADS

Both Terzaghi's and Meyerhof's methods assume the foundation is loaded centrally. Where this is not the case, the bearing pressure will vary across the base of the foundation, as illustrated in Figure 11.7. In this case the maximum pressure, q_1, should not exceed the allowable bearing capacity. Values of q_1 and q_2 can be calculated by assuming that the contact pressure varies linearly across the base, as shown in the figure. Equating vertical forces per unit length,

$$Q = \tfrac{1}{2}(q_1 + q_2)\,B$$

Taking moments about A

$$Q.d = (q_2.B \times {}^1/_2B) + {}^1/_2(q_1 - q_2).B \times {}^1/_3B$$
$$= {}^1/_2B^2({}^1/_3q_1 + {}^2/_3q_2)$$

These two equations can be used to eliminate q_2 and q_1, respectively, to give:

$$q_1 = \frac{2Q}{B^2}(2B - 3d)\ ; \qquad q_2 = \frac{2Q}{B^2}(3d - B)$$

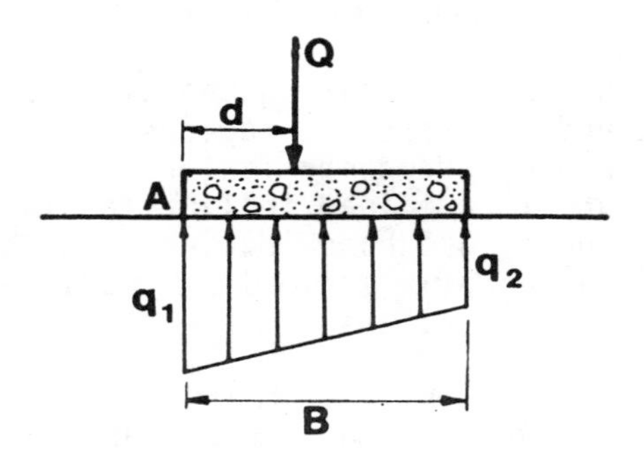

Figure 11.7 Pressures beneath a foundation where the applied load is within the middle third.

The second equation implies that for loads outside the middle third q_2 will be negative, that is, tensile. Since tensile forces cannot exist between the base and soil, we must modify the analysis for loads outside the middle third, as shown in Figure 11.8. For this case, equating vertical forces per unit length,

$$Q = \tfrac{1}{2}\, q_1 . x$$

Taking moments about A:

$$Q.d = {}^1/_2 q_1 x \times {}^1/_3 x$$
$$= {}^1/_6 q_1 x^2$$

These two equations can be used to eliminate x and q_1 respectively, to give:

$$q_1 = {}^2/_3 . Q/d \; ; \qquad x = 3d$$

For eccentric loads in two directions, on square or rectangular foundations, the analysis is first carried out as though the load were eccentric in one direction only (say the x direction shown in Figure 11.9) to give pressures q_1 and q_2. It can then be carried out in the

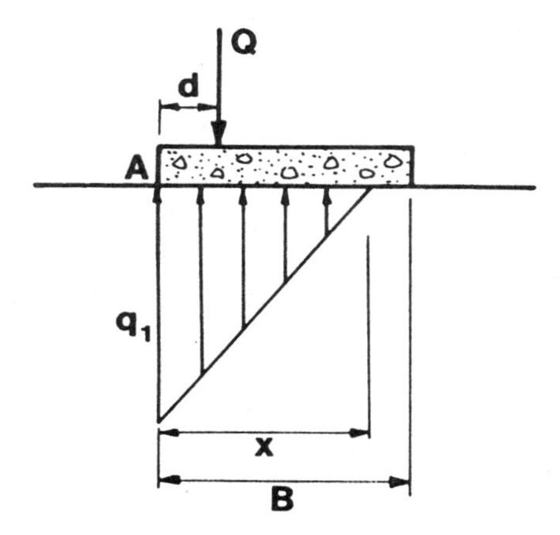

Figure 11.8 Pressures beneath a foundation where the applied load is outside the middle third.

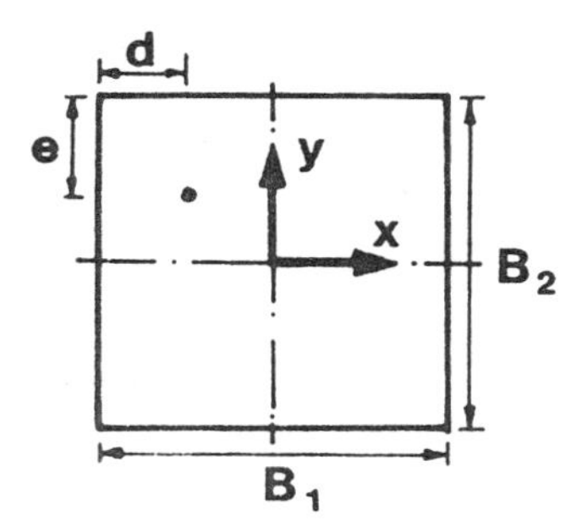

Figure 11.9 Eccentric loading on a rectangular foundation.

direction at right angles, using equivalent loads q_1B_2 for one edge and q_2B_2 for the opposite edge. This will give four values of bearing pressure, one at each corner. For loads outside the middle third, a similar procedure is adopted, using the appropriate equations.

CHANGES OF STRATA BENEATH A FOUNDATION

Where more than one type of soil lies within the potential failure zone beneath a foundation, the ultimate bearing capacity can be properly assessed only by consideration of potential failure surfaces, in a similar manner to methods used in slope stability. A number of techniques exist but the process is tedious and complicated. Usually, it is sufficient to assume that the weakest material exists throughout and to use the standard methods described in this section. In the case of a sand and a clay layer, it may be necessary to calculate the allowable bearing pressure for each material and adopt the lower value. Careful judgement is needed when dealing with complex soil conditions.

One technique which may be used where the founding stratum is underlain by a weaker stratum is to assume that the load is spread at 30°, as shown in Figure 11.10, and replace the actual foundation of width B by an effective foundation of width b at the top of the weak layer, where:

$$b = B + 2d.\tan 30^{o} \simeq B + d$$

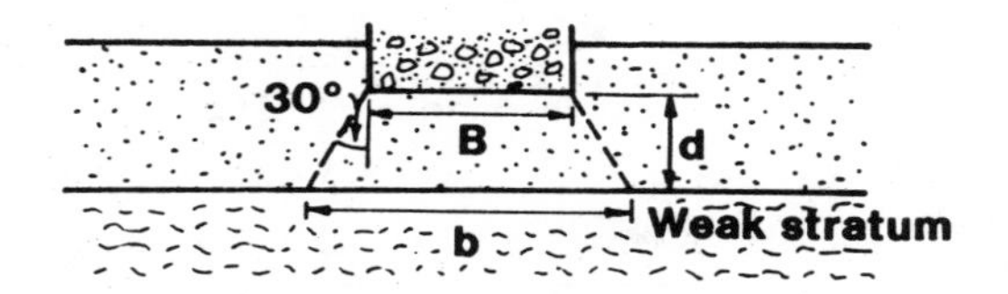

Figure 11.10 A foundation with an underlying weak stratum.

An ultimate net bearing pressure q'_{nu} for the effective foundation can then be obtained in the usual way. This will be equivalent to an ultimate bearing pressure beneath the real foundation of:

strip: $$q_{nu} = q'_{nu} \times \frac{b}{B}$$

square: $$q_{nu} = q'_{nu} \times \left(\frac{b}{B}\right)^2$$

In addition, the value of q_{nu} must not, of course, exceed the ultimate bearing capacity of the upper stratum.

Where the foundation rests on a strong stratum which is overlain by a weaker stratum (Figure 11.11), Terzaghi's equations are appropriate, so that the overlying weaker stratum is replaced by an equivalent overburden pressure and its strength is ignored.

Figure 11.11 A foundation on a strong stratum with an overlying weak stratum.

STRESSES AND DISPLACEMENTS CALCULATED BY ELASTIC THEORY

Stresses

Perhaps the first question we should consider is why we should be interested in stress distributions beneath foundations, since this information is not needed to calculate ultimate bearing capacity. A knowledge of stress distributions is useful for two reasons. Firstly, it allows us to check the depth of soil which will be noticeably affected by a foundation and to assess whether a particular stratum at depth will have any influence on its behaviour. Secondly, we need to know the stress in the soil before we can calculate settlements.

In calculating stress distributions within a soil mass it is convenient to assume linear elastic conditions. When a soil is near failure, local plastic deformations will occur and conditions will be non-linear, but if the factor of safety against shear failure is greater than about 3 then calculations based on linear elastic theory give sufficiently accurate results for most practical purposes. A large number of standard solutions exist and it is usually possible to find a solution which models the ground and loading conditions reasonably well. The most commonly used solution is probably that for a strip foundation with load p per unit area on a soil of infinite depth, shown in Figure 11.12. Table 11.1 gives solutions to the problem at various depths and distances from the centreline. From the expressions given in the figure

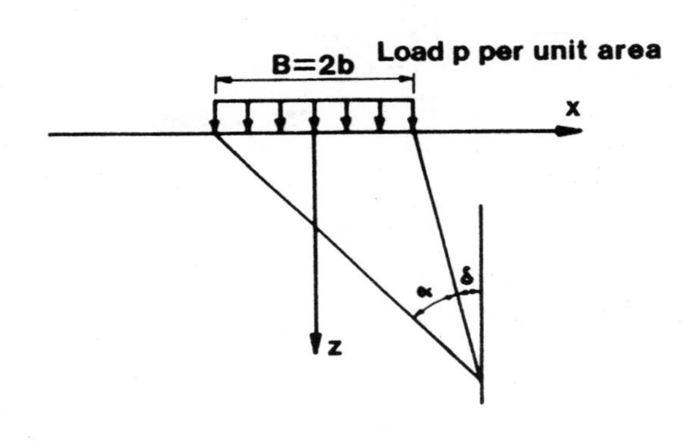

$$\sigma_z = \frac{p}{\pi}\{\alpha + \sin\alpha\cos(\alpha+2\delta)\} \qquad \sigma_1 = \frac{p}{\pi}\{\alpha + \sin\alpha\}$$

$$\sigma_x = \frac{p}{\pi}\{\alpha - \sin\alpha\cos(\alpha+2\delta)\} \qquad \sigma_3 = \frac{p}{\pi}\{\alpha - \sin\alpha\}$$

$$\sigma_y = \frac{p}{\pi}\nu\alpha \qquad \tau_{max} = \frac{p}{\pi}\sin\alpha$$

$$\tau_{xz} = \frac{p}{\pi}\sin\alpha\sin(\alpha+2\delta)$$

Figure 11.12 Stresses beneath a loaded strip on a homogeneous, isotropic elastic material of infinite depth.

it can be seen that, for a given loading, σ_1, σ_3 and τ_{max} depend only on the value of α. This means that the loci of values of constant principal stress or constant maximum shear stress lie on circles, as shown in Figure 11.13. This leads to the concept of a bulb of pressure which encloses a volume of soil that is

TABLE 11.1
STRESSES BENEATH A UNIFORMLY LOADED STRIP

x/b	z/b	σ_z/p	σ_x/p	τ_{zx}/p	β	τ_{max}/p	σ_1/p	σ_3/p
0	0	1.0000	1.0000	0	0	0	1.0000	1.0000
	.5	.9594	.4498	0	0	.2548	.9594	.4498
	1	.8183	.1817	0	0	.3183	.8183	.1817
	1.5	.6678	.0803	0	0	.2937	.6678	.0803
	2	.5508	.0410	0	0	.2546	.5508	.0410
	2.5	.4617	.0228	0	0	.2195	.4617	.0228
	3	.3954	.0138	0	0	.1908	.3954	.0138
	3.5	.3457	.0091	0	0	.1683	.3457	.0091
	4	.3050	.0061	0	0	.1499	.3050	.0061
0.5	0	1.0000	1.0000	0	0		1.0000	1.0000
	.25	.9787	.6214	.0522	$8^\circ35'$	.1871	.9871	.6129
	.5	.9028	.3920	.1274	$13^\circ17'$	.2848	.9323	.3629
	1	.7352	.1863	.1590	$14^\circ52'$	.3158	.7763	.1446
	1.5	.6078	.0994	.1275	$13^\circ18'$	.2847	.6370	.0677
	2	.5107	.0542	.0959	$11^\circ25'$	.2470	.5298	.0357
	2.5	.4372	.0334	.0721	$9^\circ49'$	.2143	.4693	.0206
1	.25	.4996	.4208	.3134	$41^\circ25'$	.3158	.7760	.1444
	.5	.4969	.3472	.2996	$37^\circ59'$	.3088	.7308	.1133
	1	.4797	.2250	.2546	$31^\circ43'$	.2847	.6371	.0677
	1.5	.4480	.1424	.2037	$26^\circ34'$	.2546	.5498	.0406
	2	.4095	.0908	.1592	$22^\circ30'$	.2251	.4751	.0249
	2.5	.3701	.0595	.1243	$19^\circ20'$	.1989	.4137	.0159
1.5	.25	.0177	.2079	.0606	$73^\circ47'$	.1128	.2281	.0025
	.5	.0892	.2850	.1466	$61^\circ50'$	.1765	.3636	.0106
	1	.2488	.2137	.2101	$47^\circ23'$	.2115	.4428	.0198
	1.5	.2704	.1807	.2022	$38^\circ44'$	.2071	.4327	.0184
	2	.2876	.1268	.1754	$32^\circ41'$	.1929	.4007	.0143
	2.5	.2851	.0892	.1469	$28^\circ09'$	.1765	.3637	.0106
2	.25	.0027	.0987	.0164	$80^\circ35'$	.0507	.1014	.0002
	.5	.0194	.1714	.0552	$71^\circ59'$	.0940	.1893	.0014
	1	.0776	.2021	.1305	$58^\circ17'$	.1424	.2834	.0052
	1.5	.1458	.1847	.1568	$48^\circ32'$	.1578	.3232	.0074
	2	.1847	.1456	.1567	$41^\circ27'$	.1579	.3232	.0073
	2.5	.2045	.1256	.1442	$36^\circ02'$	.1515	.3094	.0064
2.5	.5	.0068	.1104	.0254	$76^\circ43'$	.0569	.1141	.0003
	1	.0357	.1615	.0739	$65^\circ12'$	.0970	.1957	.0016
	1.5	.0771	.1645	.1096	$55^\circ52'$	.1180	.2388	.0029
	2	.1139	.1447	.1258	$48^\circ32'$	.1265	.2556	.0036
	2.5	.1409	.1205	.1266	$42^\circ45'$	.1269	.2575	.0036
3	.5	.0026	.0741	.0137	$79^\circ25'$	.0379	.0758	.0001
	1	.0171	.1221	.0449	$69^\circ42'$	.0690	.1384	.0005
	1.5	.0427	.1388	.0757	$61^\circ15'$	.0895	.1803	.0012
	2	.0705	.1341	.0954	$54^\circ12'$	.1006	.2029	.0018
	2.5	.0952	.1196	.1036	$48^\circ20'$	.1054	.2128	.0020
	3	.1139	.1019	.1057	$43^\circ22'$	.1058	.2137	.0020

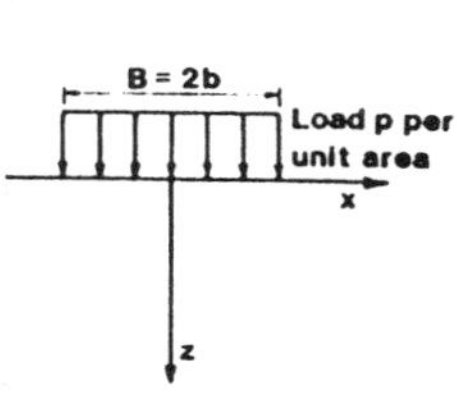

β is the angle between the direction of σ_1 and the vertical.

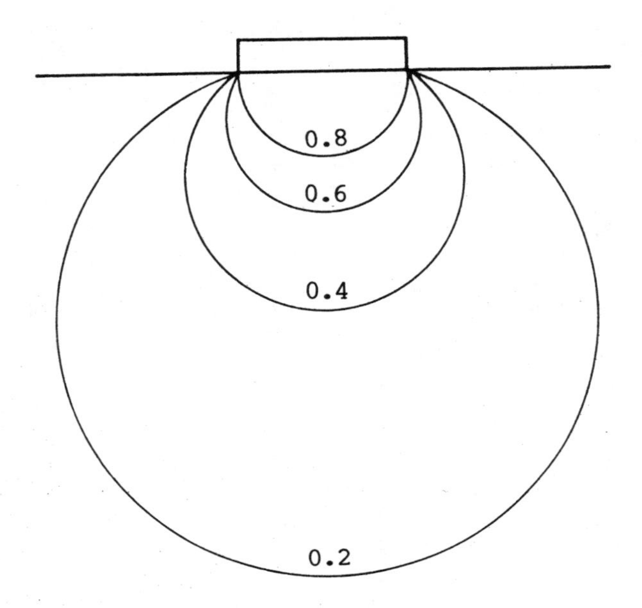

Figure 11.13 Bulb of pressure beneath a strip footing: values indicate the maximum principle stress as a proportion of the applied surface loading.

noticeably stressed by the foundation. Beyond this bulb the foundation will have negligible effect on the soil and, in the same way, it will not be affected by the properties of the soil outside the bulb. The bulb of pressure is usually considered to extend to the point where σ_1 = 0.1p to 0.2p. Further examples of bulbs of pressure are given in Figure 2.1.

Another useful standard solution is for the stress beneath the corner of a uniformly loaded area on a soil of infinite depth, as shown in Figure 11.14. Values of vertical pressure, which are of most interest when considering consolidation settlements, can be obtained from Table 11.2. Although the solution gives values of stress beneath a corner, it can be used to obtain values at any point beneath a loaded rectangular area by dividing the area into a number of smaller rectangles, as illustrated in Figure 11.15, and adding together the stresses caused by each of these.

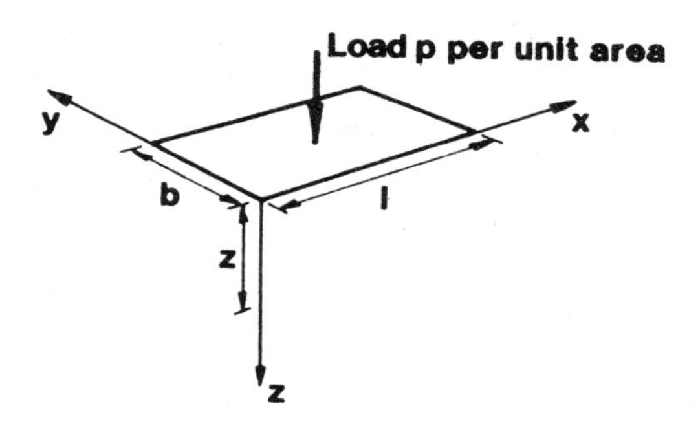

Figure 11.14 Loading on a rectangular area; for use with Table 11.2.

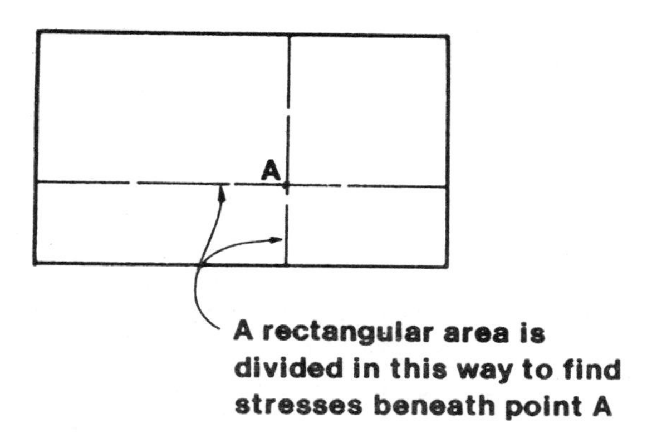

Figure 11.15 Example of how an area can be divided into sub-areas to find stresses beneath points other than the corners, using Table 11.2.

A more versatile method of calculating stresses beneath foundations of any shape is by use of Newmark charts. A large number of such charts have been developed for calculating normal and shear stresses at depth, and displacements at the surface and at depth. The chart shown in Figure 11.16 is for calculations of vertical normal stress (vertical pressure). A plan of the loaded area is drawn to scale on tracing paper. The scale must be chosen so that the basic length z

TABLE 11.2
INFLUENCE FACTORS I FOR VERTICAL PRESSURE σ_z UNDER THE CORNER OF A RECTANGULAR LOADED AREA WHERE $\sigma_z = I_p$ (SEE FIGURE 11.14 FOR DEFINITIONS)

z/ℓ \ b/ℓ	0	0.1	0.2	1/3	0.4	0.5	2/3	1	1.5	2	2.5	3	5	10	∞
0	0.000		0.250	0.250	0.250	0.250	0.250	0.250	0.250	0.250	0.250	0.250	0.250	0.250	0.250
0.2	0.000	0.137	0.204	0.234	0.240	0.244	0.247	0.249	0.249	0.249	0.249	0.249	0.249	0.249	0.249
0.4	0.000	0.076	0.136	0.187	0.202	0.218	0.231	0.240	0.243	0.244	0.244	0.244	0.244	0.244	0.244
0.5	0.000	0.061	0.113	0.164	0.181	0.200	0.218	0.232	0.238	0.239	0.240	0.240	0.240	0.240	0.240
0.6	0.000	0.051	0.096	0.143	0.161	0.182	0.204	0.223	0.231	0.233	0.234	0.234	0.234	0.234	0.234
0.8	0.000	0.037	0.071	0.111	0.127	0.148	0.173	0.200	0.214	0.218	0.219	0.220	0.220	0.220	0.220
1	0.000	0.028	0.055	0.087	0.101	0.120	0.145	0.175	0.194	0.200	0.202	0.203	0.204	0.205	0.205
1.2	0.000	0.022	0.043	0.069	0.081	0.098	0.121	0.152	0.173	0.182	0.185	0.187	0.189	0.189	0.189
1.4	0.000	0.018	0.035	0.056	0.066	0.080	0.101	0.131	0.154	0.164	0.169	0.171	0.174	0.174	0.174
1.5	0.000	0.016	0.031	0.051	0.060	0.073	0.092	0.121	0.145	0.156	0.161	0.164	0.166	0.167	0.167
1.6	0.000	0.014	0.028	0.046	0.055	0.067	0.085	0.112	0.135	0.148	0.154	0.157	0.160	0.160	0.160
1.8	0.000	0.012	0.024	0.039	0.046	0.056	0.072	0.097	0.121	0.133	0.140	0.143	0.147	0.148	0.148
2	0.000	0.010	0.020	0.033	0.039	0.048	0.061	0.084	0.107	0.120	0.127	0.131	0.136	0.137	0.137
2.5	0.000	0.007	0.013	0.022	0.027	0.033	0.043	0.060	0.080	0.093	0.101	0.106	0.113	0.115	0.115
3	0.000	0.005	0.010	0.016	0.019	0.024	0.031	0.045	0.061	0.073	0.081	0.087	0.096	0.099	0.099
4	0.000	0.003	0.006	0.009	0.011	0.014	0.019	0.027	0.038	0.048	0.055	0.060	0.071	0.076	0.076
5	0.000	0.002	0.004	0.006	0.007	0.009	0.012	0.018	0.026	0.033	0.039	0.043	0.055	0.061	0.062
10	0.000	0.000	0.001	0.002	0.002	0.002	0.003	0.005	0.007	0.009	0.011	0.013	0.020	0.028	0.032
15	0.000	0.000	0.000	0.001	0.001	0.001	0.001	0.002	0.003	0.004	0.005	0.006	0.010	0.016	0.021
20	0.000	0.000	0.000	0.000	0.000	0.001	0.001	0.001	0.002	0.002	0.003	0.004	0.006	0.010	0.016
50	0.000	0.000	0.000	0.000	0.000	0.000	0.000	0.000	0.000	0.000	0.000	0.001	0.001	0.002	0.006

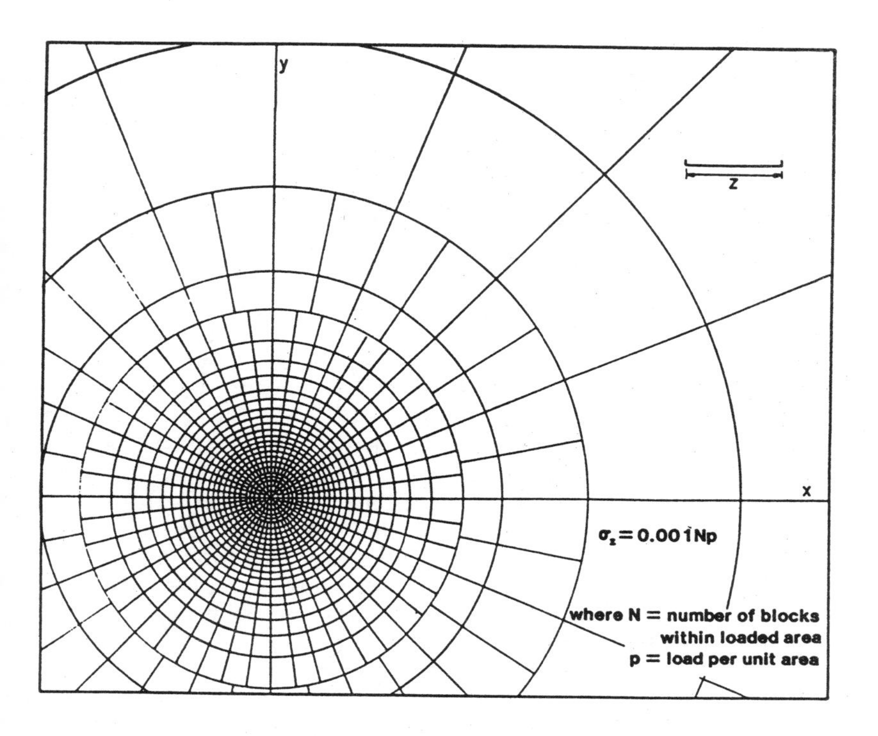

Figure 11.16 Newmark chart for vertical pressure beneath a loaded area.

marked on the chart is equal to the depth at which the value of stress is required. The scaled plan is placed on the chart so that the centre of the chart is at the point beneath which the stress value is required. The number of blocks lying within the plan is counted, estimating parts of blocks by eye, and the vertical pressure σ_z is calculated from:

$$\sigma_z = 0.001Np$$

where N is the number of blocks within the loaded area
and p is the surface load per unit area

Displacements

Elastic solutions may be used to calculate displacements directly, dispensing with the need to first calculate stresses and then obtain settlements from consolidation theory. However, this is rarely done, partly because of the need to obtain values for Young's modulus, E, and Poisson's ratio, ν, neither of which is measured in standard soil testing. Another disadvantage is that soils are not homogeneous; their consolidation properties change with depth. Further, many engineers believe that displacement calculations using elastic theory can be applied only to truly elastic movements and that they cannot be used to obtain consolidation settlements. In fact, standard elastic solutions for displacements can be applied to consolidation calculations provided the values of the elastic constants used reflect the consolidation characteristics of the soil. The main difficulty is in establishing a suitable value for Poisson's ratio, which is virtually impossible to measure. This can be overcome by turning to consolidation theory, which ignores lateral strains induced during consolidation: equivalent to a Poisson's ratio of zero. There are two reasons for thinking that this is a valid assumption: first, that lateral strains during consolidation (reported by Skempton and others) are usually small; and second, simply that consolidation theory has been found to give reasonably accurate results when full allowance is made for dissipation of the excess pore water pressure caused by the consolidation loading. A suitable value of Young's modulus for consolidation calculations may be obtained from the value of the coefficient of volume compressibility, m_v, measured in standard laboratory consolidation tests, using the relationship

$$E = \frac{1}{m_v} \cdot \frac{(1+\nu)(1-2\nu)}{(1-\nu)} = \frac{1}{m_v} \quad \text{if } \nu = 0$$

Surface displacements caused by loading from a foundation of any shape may be obtained from the Newmark chart given in Figure 11.17. This is used in a similar manner to the Newmark chart for vertical stresses, except that the plan of the loaded area may

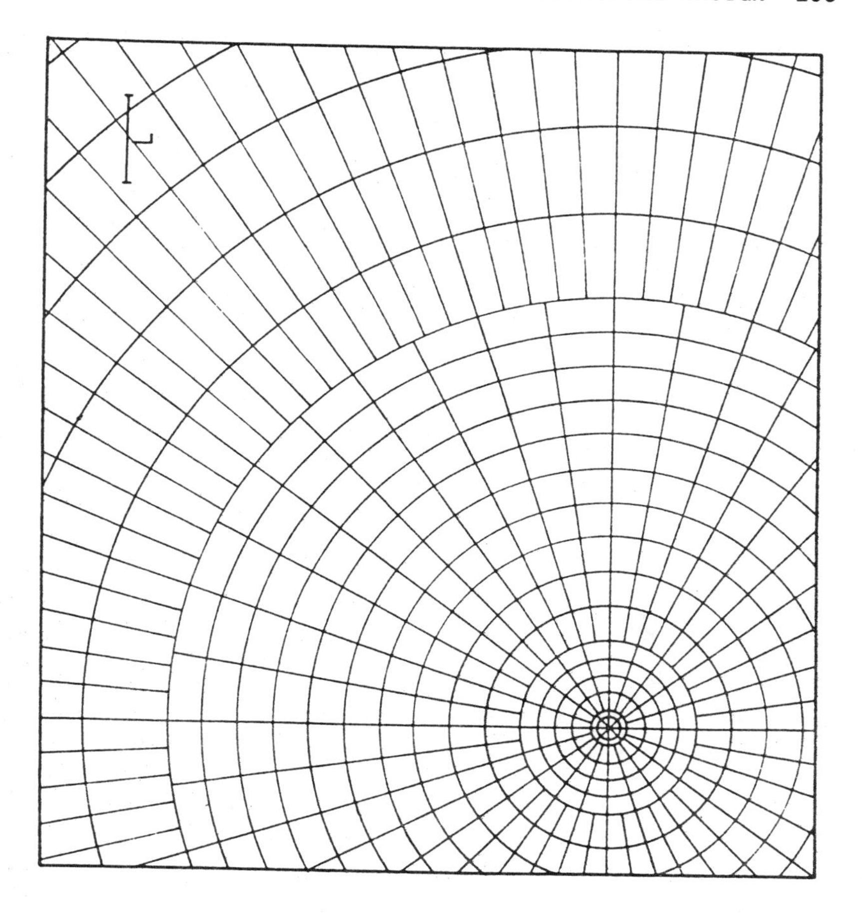

$$\rho_0 = 0.02(1-\nu^2)\frac{pL}{E}N$$

where N = number of blocks within loaded area
p = load per unit area
L = scaled distance of basic length line marked L on chart
E = Young's modulus
ν = Poisson's ratio

Figure 11.17 Newmark chart for vertical displacement beneath a loaded area.

be drawn to any appropriate scale and the surface displacement at the centre of the chart is obtained from

$$p_o = 0.02(1-\nu^2).\ \frac{pL}{E}.\ N$$

where p is the load per unit area of the loaded area
L is the scaled distance represented by the length L marked on the chart
N is the number of blocks within the loaded area
ν and E are the elastic constants discussed above.

CONSOLIDATION SETTLEMENTS

Total settlement

The coefficient of volume compressibility, m_v, was defined in Chapter 10 as:

$$m_v = \frac{1}{h} \cdot \frac{dh}{dp}$$

where h is the specimen thickness in a consolidation test
and dh is the change in thickness due to a pressure change dp

This definition can be used to calculate settlement due to a compressible layer beneath a foundation, as illustrated in Figure 11.18(a). Specimen thickness h is replaced by the layer thickness H and the change in thickness dh becomes the settlement ρ. The stress increase dp is now equal to the vertical pressure in the layer, σ_z, due to the applied loading. Substituting these terms and rearranging the equation gives:

$$\rho = H.m_v.\sigma_z$$

The value of σ_z is usually obtained from elastic analysis, as described earlier. The stress across the layer will obviously vary with depth but for thin layers it is usually sufficient to calculate the value

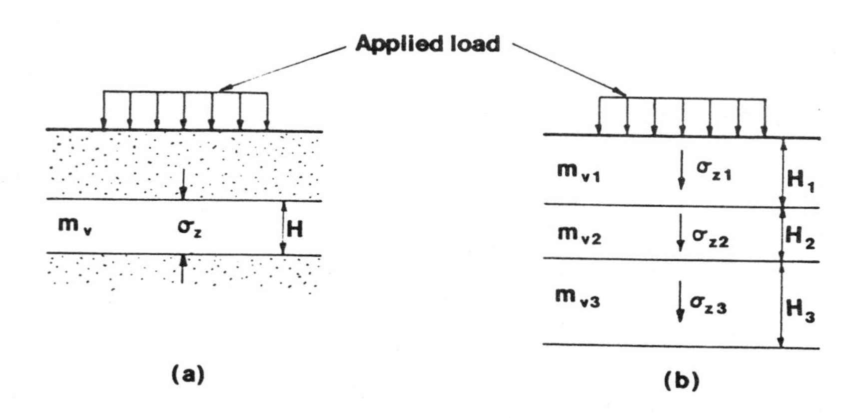

Figure 11.18 Consolidation settlement due to compressible soil: (a) a single compressible layer; (b) multiple layers.

of σ_z at mid-depth. Alternatively values at the top and bottom of the layer may be calculated and the average value used. If several compressible layers are present, as illustrated in Figure 11.18(b), then values of σ_z are calculated for each of these and the settlement is obtained by adding together the settlements of all the layers:

$$\rho = H_1 m_{v1} \sigma_{z1} + H_2 m_{v2} \sigma_{z2} + \ldots$$

This approach can also be used with thick layers of material to overcome the problem that σ_z will vary appreciably over the depth. The layer is arbitrarily divided into a number of sub-layers and average values of σ_z are calculated for each of these layers. The values of m_v can also be adjusted for each layer to take into account the variations caused by the different amounts of overburden on each layer.

The settlement calculations described above can be speeded up by the use of the influence factors given in Figure 11.19. The derivation of these can be appreciated by considering the settlement, dρ, of a thin layer of soil, thickness dz, at depth z beneath a foundation, as shown in Figure 11.19. From the above expressions for settlement,

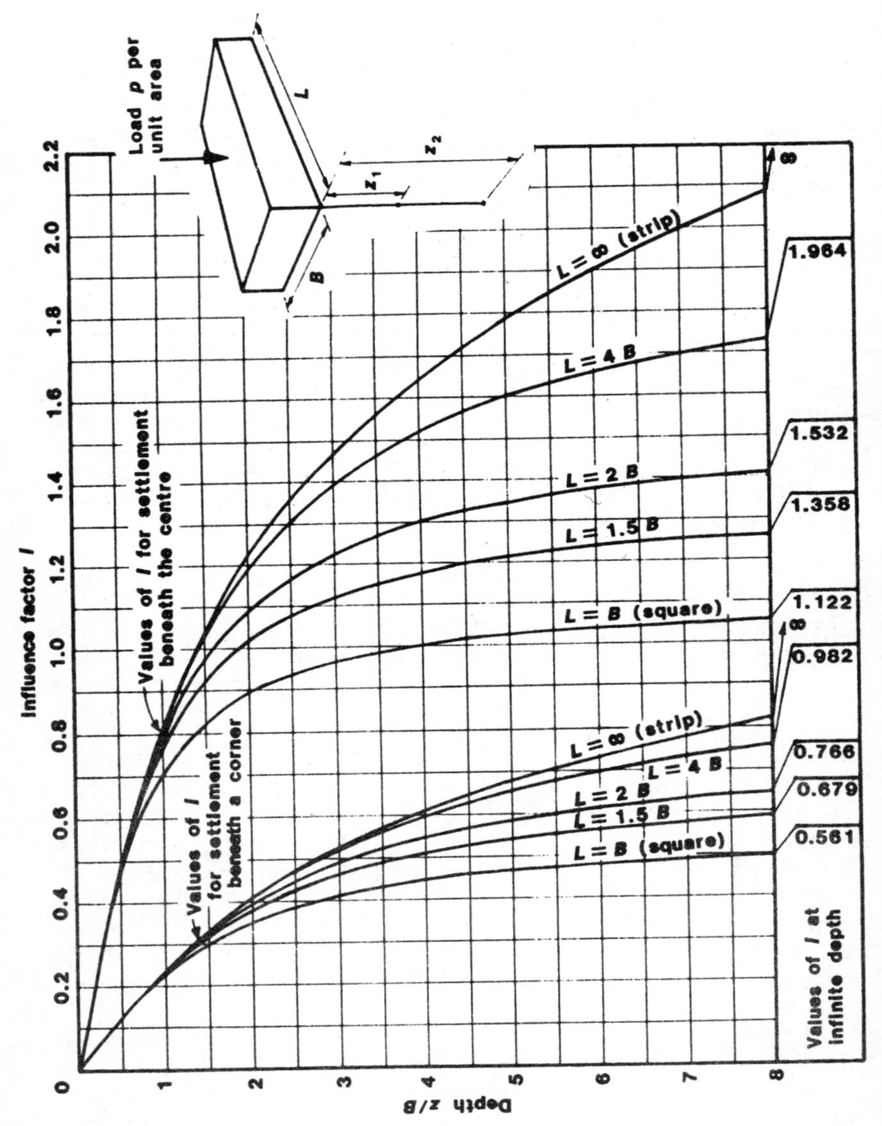

Figure 11.19 Influence factors $I = \frac{1}{Bp}\int \sigma_z.dz$ *for the calculation of settlement* ρ *within a layer of thickness z, beneath the corner and centre of a rectangular foundation, where* $\rho = m_v.B.p.I$.

$$d\rho = m_v.\sigma_z.dz$$

where σ_z is the vertical pressure at depth z due to the applied loading.

For a whole layer from depth z_1 to depth z_2, settlement will be

$$\rho = m_v\int_{z_1}^{z_2} \sigma_z.dz = m_v[\int_0^{z_2} \sigma_z.dz - \int_0^{z_1} \sigma_z.dz]$$

assuming m_v to be constant throughout the layer. For a given foundation shape (L/B ratio), σ_z depends on the applied loading and on depth expressed in terms of footing widths (i.e. on the z/B or z/l ratio - see Table 11.2 for illustration of this), so the integral will have a fixed value for a given applied pressure p and z/B ratio. Thus, we can obtain an influence factor I:

$$I = \int\frac{1}{Bp}\int_0^z \sigma_z.dz$$

which is independent of loading and footing width. The expression for settlement then becomes

$$\rho = m_v.B.p(I_2 - I_1)$$

where I_1 and I_2 are values of the influence factor at depths z_1 and z_2, respectively. Influence factors for both the centre and corner of a rectangle are given in Figure 11.19. Settlement may thus be calculated beneath the centre and corner of a rectangular foundation for compressible layers of any thickness and depth, without the need to evaluate stresses. Where m_v varies with depth, the profile may be divided into several layers and settlement obtained as the sum of the individual layers.

Stress and displacement calculations are usually based on the assumption that the loaded area is completely flexible, resulting in settlements at the centre of the area being greater than those at the edges: this point is particularly brought out by the influence charts of Figure 11.19. This assumption is obviously not true where loading is through a rigid foundation, which must settle as a solid body. The settlement of rigid foundations can be obtained approximately from the following expressions:

circle or strip, $\rho_{rigid} = {}^1/_2(\rho_{centre} + \rho_{edge})_{flexible}$

rectangle, $\rho_{rigid} = {}^1/_3(2.\rho_{centre} + \rho_{corner})_{flexible}$

Corrections to settlement calculations

In many soils calculated settlement values overestimate the amount of consolidation settlement. A correction is applied to give a more accurate estimate of the true consolidation settlement, thus:

$$\rho_{field} = \mu.\rho$$

where ρ_{field} is the estimated settlement taking into account the differing pore pressure response of the soil in the field to that in the oedometer test,

ρ is the settlement estimate obtained by the simple application of consolidation theory, and

μ is a factor which depends on the soil type and thickness of the compressible layer compared with the foundation width.

The shape of the foundation has little effect on the value of μ. For most practical purposes values of μ obtained from Table 11.3 may be used.

Differential settlement

Even for completely homogeneous conditions, parts of a loaded rectangular area settle more than others; the centre experiencing the most settlement and the corners the least. In general, settlements tend to be greatest at the centres of structures and least at their edges. This differential settlement causes distortion of structures and is often more important than total settlement because of its damaging effects. The amount of differential settlement caused in this way can be estimated by elastic analysis, although the calculations can often be tedious.

A second cause of differential settlement is

TABLE 11.3
VALUES OF CONSOLIDATION FACTOR μ FOR VARIOUS SOIL TYPES

Type of clay	μ			Definitions of H and b	
	H/b=0.5	H/b=1	H/b=4		
Very sensitive clays (soft alluvial, estuarine, marine clays)	1.0-1.1	1.0-1.1	1.0-1.1	B; H; Compressible layer Surface layer	B; 60°; z; Assumed spread of load; H; b; Compressible layer Approximate approach for subsurface layer $b = B + 2z.\cot 60^{\circ}$ $= B + z$
Normally consolidated clays	0.8-1.0	0.7-1.0	0.7-1.0		
Over-consolidated clays (Lias, London, Oxford, Weald clays)	0.6-0.8	0.5-0.7	0.4-0.7		
Heavily over-consolidated clays (Boulder clay, marl)	0.5-0.6	0.4-0.5	0.2-0.4		

variation of the ground itself. This problem should become apparent from m_v values, which may show a wide scatter in variable ground conditions. The effects of these variations can be estimated by choosing the highest and lowest likely values of m_v to obtain maximum and minimum values for settlement.

It is impossible to make accurate predictions of differential settlement and careful judgement is needed to avoid improbable overestimates or underestimates. For most simple designs it is usually assumed that differential settlement will be half the maximum total settlement.

Rate of settlement

As discussed in Chapter 10, the rate at which settlement occurs is defined by the formula:

$$t = \frac{T_v d^2}{c_v}$$

where t = the time for a required degree of settlement
T_v = basic time factor

d = maximum length of the drainage path
c_v = coefficient of consolidation

The time required for settlement to be substantially (i.e. 90%) complete is obtained for practical purposes by using a value of basic time factor $T_v = 0.8$. More precise values of T_v for various degrees of consolidation are given in Table 10.1. The maximum length of the drainage path, d, is equal to the thickness of the consolidating layer if drainage is from the top or bottom only; for drainage from both top and bottom, d equals half the layer thickness. With more complicated soil conditions, where there are sand or silt layers which may or may not form continuous drainage paths and where lateral drainage may be significant, the value of d may be difficult to assess. Because time t is proportional to d^2 then, unless drainage conditions can be fairly accurately determined, there is little hope of predicting the rate of consolidation with any reasonable accuracy.

Quick settlement estimates for simple foundations

A glance at the influence factor curves given in Figure 11.19 shows that when a foundation settles the majority of the compression takes place within the upper layers of soil, down to about three times the footing width for square foundations and six times the footing width for strip foundations. It is interesting to note that, in normal foundation practice, rectangular foundations are assumed to influence the soil to a depth of 1.5 foundation widths. Whilst the majority of settlement does take place within this depth, Figure 11.19 indicates that some measurable compression of the soil down to the depths given above can take place. Referring back to Figure 11.13, it can also be seen that in terms of stresses the depth of soil influenced by a foundation is greater than the 1.5 foundation widths often assumed. Allowing for the fact that soils also generally become stiffer with depth, most of the compression can usually be expected to take place within a depth equal to about three footing widths below the base of a foundation. This gives influence factor values of 0.41 and 0.96 for settlement beneath the corner and centre, respectively, of a square foundation. Thus, for a square foundation on a homogeneous soil:

settlement at corner = $0.41\ m_v.B.p$
settlement at centre = $0.96\ m_v.B.p$

Applying the approximate relationship between settlements of rigid and flexible foundations, given earlier, the settlement, ρ, of a rigid square footing would be

$$\rho_{square} = {}^1/_3(0.96 \times 2 + 0.41)m_v.B.p$$

Thus, $\rho_{square} \approx 0.8\ m_v.B.p$

A similar exercise for strip foundations gives influence factors of 0.52 (corner) and 1.46 (centre), giving

$$\rho_{strip} = {}^1/_3(1.46 \times 2 + 0.52)$$

Thus, $\rho_{strip} \approx 1.1\ m_v.B.p$

Settlements on sands

In sands and gravels the allowable bearing pressure is usually limited by settlement considerations rather than by ultimate bearing capacity. For most ground investigations it is impractical to obtain and test undisturbed samples of granular soils. The most common procedure is to estimate the relative density from standard penetration tests or cone penetrometer tests. Settlement can be estimated from these values using the chart given in Figure 11.20. This gives the allowable bearing pressure for 25mm settlement. If the acceptable settlement is other than 25mm, the allowable bearing pressure can be altered on a pro-rata basis, provided it is within acceptable limits based on ultimate bearing capacity. The values obtained from the chart are for dry conditions: if the water table is one foundation width below the underside of the foundation, or higher, then bearing pressures should be halved. Alternatively, the bearing pressure read from the chart will produce 50mm settlement. Large rigid foundations such as rigid rafts or deep pier foundations suffer less from settlements than other

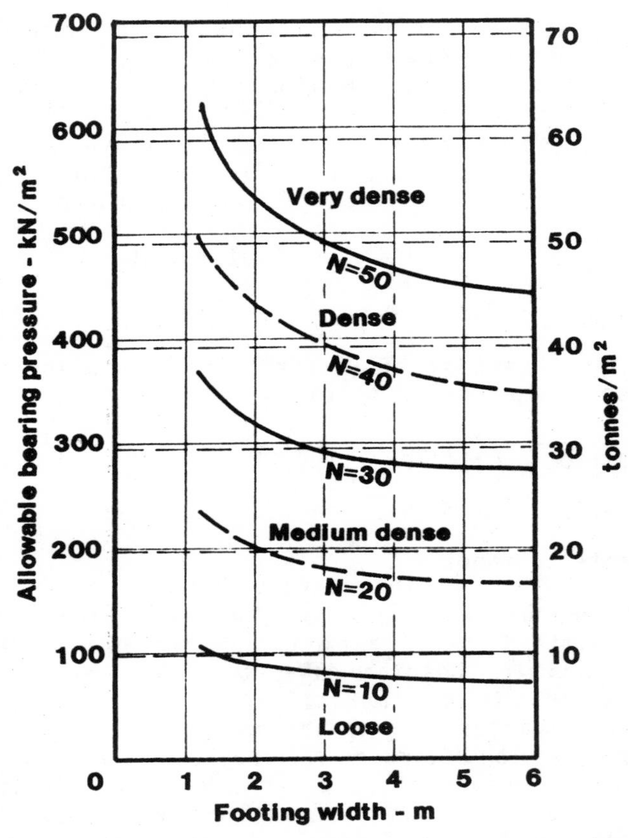

Figure 11.20 Chart for estimating allowable bearing pressure on sand using standard penetration test results. Continuous lines are based on the original chart by Terzaghi and Peck; broken lines are interpolations.

types and, for these, allowable bearing pressures may be doubled. Differential settlement is usually assumed to be half total settlement. Settlement estimates from this chart cannot be regarded as more than a rough guide and they are known to give conservative results. Some authors (e.g. Bowles, 1982) now suggest bearing pressures of about 1.5 times those obtained from Figure 11.20.

Tolerable differential settlements

The limiting factor is usually the amount of angular distortion that a structure, or any installed equipment, can accommodate without distress or damage. It can be expressed as the angle of the settlement profile, as illustrated in Figure 11.21. Differential settlement is thus given by:

$$\Delta\rho = \tfrac{1}{2}\ \text{span} \times \tan\beta$$
$$\simeq \tfrac{1}{2}\ \text{span} \times \beta$$

(since for small angles, $\tan\beta \simeq \beta$ where β is in radians)

Typical values of tolerable differential settlement are given in Table 11.4.

PILED FOUNDATIONS

The design of piled foundations is a specialist topic about which many books have been written. The information given in this section shows how the load capacity of piles is calculated for the simple case of vertical loads on straight-sided piles but it is not intended to equip the reader to carry out his own pile designs. These are usually carried out by piling specialists who work with a specialist piling contractor.

The first step is to calculate the carrying capacity of a single pile. A suitable factor of safety is then chosen to obtain an allowable working load. The carrying capacity of a group of closely-spaced piles may not be the same as the sum of the carrying capacities of all the piles in the group so an allowance must sometimes be made for the group effects. Another problem that can arise in certain soil conditions is that the soil surrounding the upper part of the pile settles relative to the pile. This reverses the friction on the sides of the pile. This phenomenon, known as negative skin friction, produces an added load on the pile so reducing its carrying capacity. These points are discussed below.

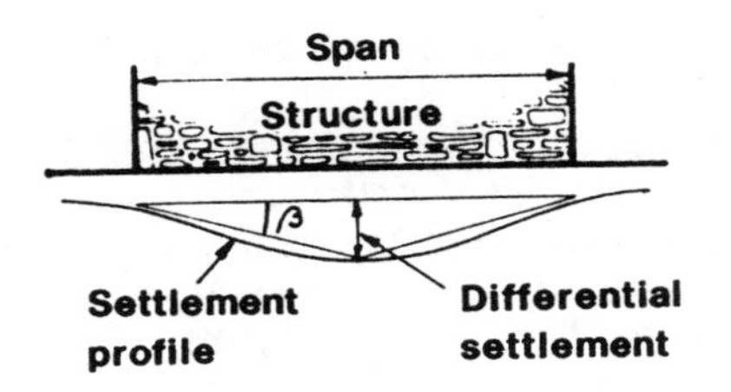

Figure 11.21 Definition of angle β in Table 11.4

TABLE 11.4
TYPICAL VALUES OF TOLERABLE DIFFERENTIAL SETTLEMENT

Type of structure	Tolerable differential settlement, β (radians)	Comments
Circular steel petrol or fluid storage tanks: fixed top floating top	0.008 0.002 - 0.003	For floating top, value depends on details of top. Values apply to tanks on a flexible base. With rigid base slabs, such settlement will cause cracking and local buckling.
Tracks for overhead travelling crane.	0.003	Value taken longitudinally along track. Settlement between tracks is not usually the controlling factor.
Rigid circular ring or mat footing for stacks, silos, water tanks etc.	0.002	
Jointed rigid concrete pressure pipe.	0.015	Value is allowable angle change at joint. This is usually 2-4 times average slope of settlement profile. Damage to joint also depends on longitudinal extension.
One- or two-storey steel framed warehouse with truss roof and flexible cladding.	0.006 - 0.008	Overhead crane, pipes, machinery or vehicles may limit tolerable values to less than this.
One- or two-storey houses or similar buildings with brick load-bearing walls.	0.002 - 0.003	Larger value is tolerable if most settlement has taken place before finishes are completed.
Structures with sensitive interior or exterior finishes such as plaster, ornamental stone or tiles.	0.001 - 0.002	
Multi-storey heavy concrete rigid framed structures on thick structural raft foundations.	0.0015	Damage to interior or exterior finish may limit value.

The ultimate load capacity of a pile in clay

The ultimate carrying capacity Q_μ is made up of adhesion Q_s and end bearing Q_b. Adhesion (often called skin friction) is usually much greater than end bearing in clays. Thus:

$$Q_u = Q_s + Q_b$$

The adhesion on a pile is given by:

$$Q_s = \alpha.\bar{c}.A_s$$

where A_s is the embedded surface area of the pile.
$\bar{c}$ is the average undrained shear strength of the clay along the sides of the pile, and
α is an adhesion factor.

Researchers have found that the value of α can vary widely so that it is difficult to allocate a value to it. For driven piles, values obtained by Nordlund, given in Figure 11.22, are usually used. For bored piles a value of 0.45 is often used but the value chosen depends on the type of clay. The value of α is generally less for harder clays: a value of 0.3 is often used in stiff fissured clays, and for hard Lias clays α may be as low as 0.1.

The end bearing is obtained from Meyerhof's equation for the bearing capacity of cohesive soils:

$$Q_b = c.N_c.A_b$$

where c is the undisturbed shear strength at the base of the pile
A_b is the pile base area (thus for a circular pile, radius R, $A_b = \pi R^2$), and
N_c is Meyerhof's bearing capacity factor, usually taken as 9.

The ultimate capacity of a pile in sand or gravel

As with clay soils, the ultimate carrying capacity is made up of end bearing and skin friction but in

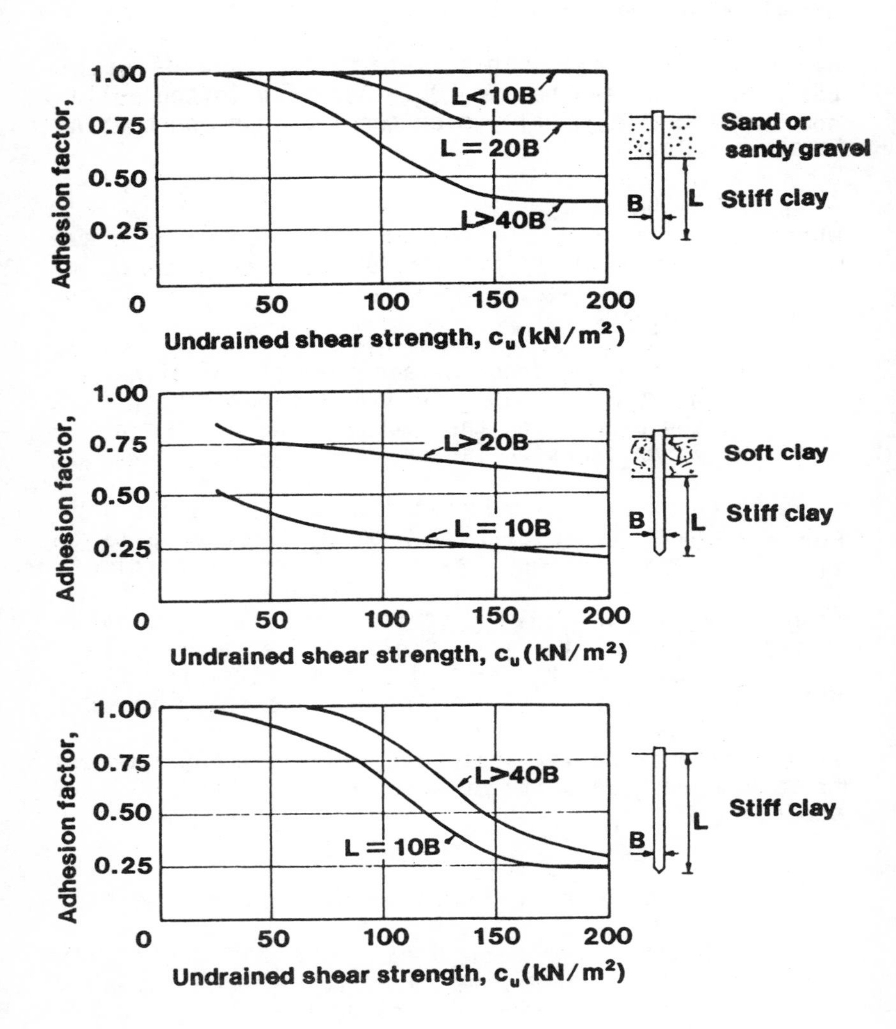

Figure 11.22 Nordlund's adhesion factors for driven piles.

granular soils most of the load is usually carried as end bearing. The skin friction f on a unit area depends on the pressure of the surrounding soil on the pile and on the coefficient of friction at the soil/pile interface. This is usually expressed as:

$$f = K_s.p_d.\tan\delta$$

where K_s is the coefficient of earth pressure on the pile shaft; the ratio of the lateral to vertical earth pressure at the sides of the pile

p_d is the overburden pressure at depth z. Generally $p_d = \Sigma\gamma z$ where γ is the bulk density for strata above the water table and the submerged density below the water table

δ is the angle of friction between the pile and the soil.

For a pile surrounded by granular soil between depths z_1 and z_2, the total skin friction is:

$$Q_s = \tfrac{1}{2}K_s\gamma(z_1 + z_2)\ \tan\delta.A_s$$

where A_s is the embedded area from z_1 to z_2 (thus, for a circular pile, radius R, $A_s = 2\pi R(z_2 - z_1)$.

If the pile is partly submerged then the contributions from above and below the water table must be calculated separately.

TABLE 11.5
VALUES OF K_s AND δ FOR DRIVEN PILES

Pile material	δ	K_s	
		Low rel. density ($\phi < 35^o$)	High rel. density ($\phi > 35^o$)
Steel	20°	0.5	1.0
Concrete	$3/4\ \phi$	1.0	2.0
Wood	$2/3\ \phi$	1.5	4.0

Values of K_s and δ, obtained by Broms, are given in Table 11.5. This is valid up to a skin friction value, f, of 110kN/m^2, which is the maximum skin friction value that should be used for any straight-sided pile.

In calculating end resistance, the third term in Meyerhof's equation (see earlier), relating to base friction, is relatively small for long slender piles and is usually ignored. Thus, base resistance is given by:

$$Q_b = p_o(N_q - 1).A_b$$

where p_o is the effective overburden pressure at the pile base, and
A_b is the area of the pile base (thus, for a circular pile of a radius R, $A_b = \pi R^2$).

Bearing capacity factor N_q depends on the value of the angle of shearing resistance ϕ for the soil which is usually obtained from standard penetration test results, using the correlations given in Figure 11.2. However, Meyerhof's values of N_q are unrealistically high for piled foundations and it is more usual to use those obtained by Berezantsev, shown in Figure 11.23.

When calculating both skin friction and end

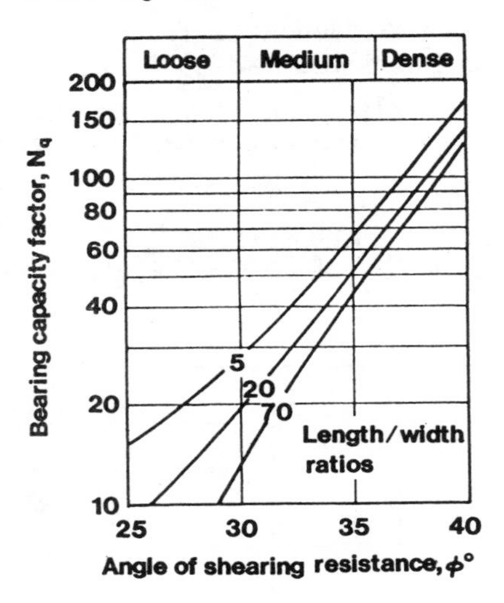

Figure 11.23 Berezantsev's bearing capacity factor, N_q.

resistance of bored piles in granular soil, a low relative density should always be assumed, whatever the initial state of the soil, to allow for disturbance during boring.

The allowable working load on a pile

The allowable working load is obtained by dividing the ultimate load by a factor of safety. Factors of safety can be applied to the ultimate load capacity or to the skin friction or end bearing separately. The allowable working load is often taken as the lower of:

$$\frac{Q_s + Q_b}{2.5} \qquad (1)$$

and

$$\frac{Q_s}{1.5} + \frac{Q_b}{3.0} \qquad (2)$$

where Q_s and Q_b are the ultimate loads carried by skin friction and end bearing, respectively. The value of Q_s in equation (1) is usually based on adhesion factors using average values of shear strength. Where the lower factor of safety is used in equation (2), adhesion factors are usually based on the lower range of shear strength values obtained.

Where piles are driven to refusal in rock or in very dense sands or gravels, the maximum allowable load is usually limited by the strength of the pile rather than the support of the soil.

Where piles pass through bands of different materials, the skin friction may be calculated for each band and the total skin friction taken as the sum of these values, unless very compressible layers are present. When calculating end bearing, care must be taken to check that weak material is not likely to occur near the tip, which would result in a decrease in end bearing capacity. If this is a possibility, it must be allowed for (by using a reduced value of N_c or N_q) when calculating end bearing. In c-ϕ soils, skin friction may be taken as the sum of friction and adhesion and end bearing may be taken as the sum of end bearing due to both cohesion and internal friction. However, results should be viewed with caution because very little information is available about the behaviour of piles in c-ϕ soils.

For a pile group, the carrying capacity of the group as a whole must also be checked.

The bearing capacity of pile groups

The failure load, Q_u, of a pile group is expressed as

$$Q_u = n.Q_{us}.E_f$$

where n is the number of piles in the group,
Q_{us} is the failure load of a single comparable pile, and
E_f is the group efficiency ratio

From this, the group efficiency ratio can be defined as

$$E_f = \frac{Q_{u/n}}{Q_{us}} = \frac{\text{average load on a pile in a group at failure}}{\text{failure load of a single comparable pile}}$$

In granular soils, driven piles compact the surrounding soil, increasing its bearing capacity, and model tests have shown that group efficiency ratios of driven piles in sand can be as high as 2. With bored piles, the action of boring tends to reduce rather than increase compaction so that the group efficiency ratio of bored pile groups is unlikely to exceed 1. For design purposes, a group efficiency ratio of 1 is typically used for all kinds of piles in granular soils even though this approach is conservative for driven piles in loose sands. That is, the effects of the group are ignored when predicting bearing capacities. However, bored piles should not be spaced closer than 3 diameters (centre to centre).

In cohesive soils, the group (both piles and contained soil) can be thought of as a deep foundation and the failure load of the group as a whole (block failure) can be calculated using Meyerhof's equation, as described earlier in this chapter. Thus, for a group of width B_1, length B_2, depth D, the failure load is

$$Q_u = B_1 B_2 c N_c \lambda + 2(B_1+B_2)D\bar{c}$$

where

N_c is Meyerhof's bearing capacity factor,
λ is a shape factor (typically 1.2 to 1.3 for piled foundations),
c is the cohesion of the clay at the base of the piles, and
$\bar{c}$ is the average cohesion of the clay over depth D.

Alternatively, the group efficiency ratio can be obtained by the graph shown in Figure 11.24 which was obtained by Whitaker from model tests.

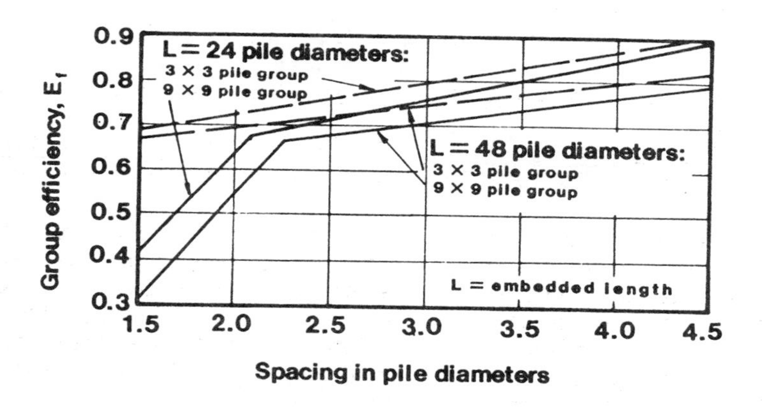

Figure 11.24 Group efficiencies for pile groups in cohesive soil.

If weak clay layers exist beneath a pile group, overstressing of the clay can occur. It may be necessary to check this.

As far as possible, all piles on the same pile cap should be of approximately equal length.

Where piles of different lengths are near to each other, it is usual to make the shorter pile of sufficient length to ensure that a line inclined at 45^0 from its base does not interesect the longer adjacent pile. This is to avoid load from the toe of the shorter pile being transmitted through the soil to the longer pile.

NEGATIVE SKIN FRICTION

When a pile is embedded in a stratum that is overlain by a more compressible clay or by fill, settlement of the overlying material may lead to a downward movement of the soil relative to the pile, resulting in a reversal of the frictional forces on the sides of the pile. Thus, far from supporting the pile, the upper layer of material is actually dragging it down, increasing the load on the upper part of the pile. The circumstances in which negative skin friction may occur can be divided into three cases:

Case 1. A pile driven through a soft sensitive clay into a relatively incompressible stratum. Remoulding of the clay during driving may be sufficient to cause settlement. Soft marine or estuarine clays may be particularly susceptible.

Case 2. A pile driven through a soft insensitive clay into a relatively incompressible stratum, with a surcharge on the surface. Normally, a soft insensitive clay overlying a stiffer clay should not cause problems but loading at the surface will lead to settlement and could give rise to negative friction on the pile. Drainage of swampy or boggy areas would have a similar effect.

Case 3. A pile driven through recently placed fill into either a compressible or a relatively incompressible stratum. Negative friction will arise as a result of consolidation of the fill. In older fills, negative friction may be lessened or absent altogether.

With the present state of knowledge it is impossible to accurately predict the extent of negative friction which will develop on a pile. However, the maximum downdrag that could develop can be estimated by assuming the distribution of frictional stresses shown in Figure 11.25.

The effects of negative skin friction can be reduced or eliminated by sleeving the pile section within the overlying fill or soft clay. Alternatively, it may be surrounded by a plastic membrane with a low friction value or coated with bitumen. If bitumen is used, care must be taken to control the application to ensure that the coating adheres properly to the pile and is at least 3mm thick.

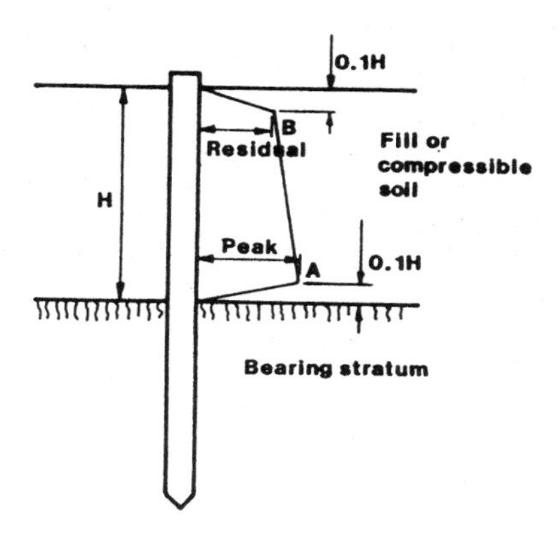

Figure 11.25 Assumed distribution of negative skin friction (modified after Tomlinson, 1980).

PILE DRIVING FORMULAS

Pile driving formulas attempt to relate the ultimate bearing capacity of a pile to its driving resistance. They all make sweeping simplifications about the nature of the soil and pile, the relationship between the static and dynamic resistance of the ground, and the way in which the load is transferred from the hammer to the pile and thence to the soil. A large number of pile driving formulas have been developed in an attempt to overcome the shortcomings of the technique but generally their use in predicting the bearing capacity of piles has proved less than satisfactory. Nevertheless, they are used extensively as an aid when driving and their usefulness can be increased by comparing the predicted bearing capacity, using a pile driving formula, with the results of pile loading tests for a particular site, and modifying the formula to bring the predicted results into line with actual results for that particular site.

Two commonly-used formulas are developed below.

The Engineering News formula

The method assumes that the hammer and pile can be treated as impinging particles and that all the hammer energy is transferred to the pile to overcome the resistance of the ground. Further, as the hammer strikes the pile, resistance of the pile increases in an elastic manner to a maximum value R and then remains constant with displacement until it finally rebounds elastically at the end of the blow, as illustrated in Figure 11.26.

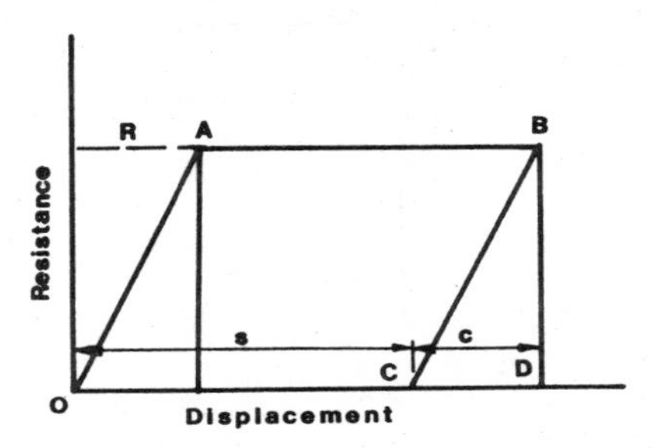

Figure 11.26 Assumed displacement/resistance relationship in the Engineering News formula.

The energy supplied by the hammer is represented by area OABD but the energy represented by triangle BCD is dissipated as the pile rebounds, leaving only the energy represented by area OABC to be usefully employed in driving the pile. Thus, for a hammer of weight W falling through height H,

Total work done = OABD = OABC + BCD

giving $$WH = R(s + \tfrac{1}{2}c)$$

where

R, the maximum resistance, represents the ultimate bearing capacity of the pile,
s is the movement per hammer blow - the "set",
c is the elastic rebound.

A formula using this reasoning was first published by A.M. Wellington in the "Engineering News" in 1888: hence it is usually referred to as the "Engineering News" formula. Empirical values are given to the constant ½c. Traditionally, values have been obtained for values of H and s measured in inches. Thus,

for drop hammers $WH = R(s + 1.0)$

for single-acting steam hammers $WH = R(s + 0.1)$

If distances are measured in millimetres, the constant ½c should be multiplied by 25. Variations of this formula are popular in North America.

The Hiley Formula

The Hiley formula takes into account energy losses occurring in the hammer system, at the point of impact and due to compression of the head assembly (dolly, helmet, packing), the pile and the surrounding ground. The energy considerations are as follows.

a. The energy at impact from a weight W falling a distance H is kWH where k is a constant which takes into account frictional loss in the hammer system.
b. The energy required to drive the pile a distance s against a driving resistance R is R.s.
c. The energy loss at impact between the hammer and the pile head assembly is

$$\frac{kWHP(1 - e^2)}{(W + P)}$$

where P is the weight of the pile and
e is the coefficient of restitution.

d. The energy loss due to elastic compressions c_c, c_p and c_q in the pile head assembly, pile, and ground, respectively, is $\frac{1}{2}R(c_c + c_p + c_q)$

Thus, the energy equation is

$$kWH = \frac{kWHP(1 - e^2)}{(W + P)} + R.s + \tfrac{1}{2}R(c_c + c_p + c_q)$$

Re-arranging gives the more usual form of the formula:

$$R = \frac{\eta . WH}{s + \frac{1}{2}c} \quad \text{where} \quad \eta = \frac{k(W + e^2P)}{(W + P)}$$

and $c = c_c + c_p + c_q$

The rebound of the pile and ground $(c_p + c_q)$ can be measured by fixing a straight edge, anchored to the ground by stakes, alongside the pile and moving a pencil slowly along it during driving so that it marks a piece of card which is attached to the pile. From the plot (Figure 11.27) the value of $(c_p + c_q)$ can be directly measured.

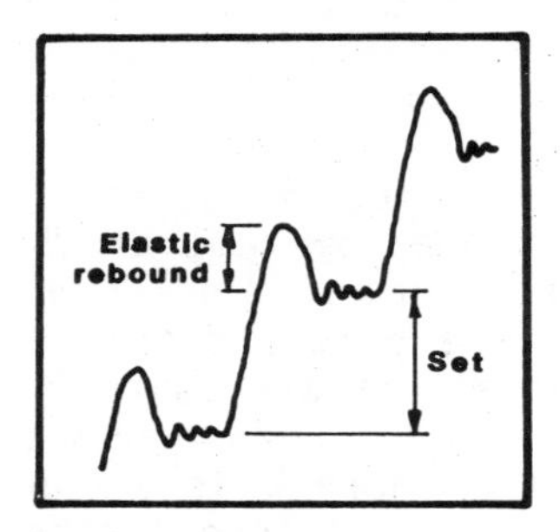

Figure 11.27 Measurement of the elastic rebound of a pile.

Set s is additionally measured at some convenient point on the rig throughout the driving procedure.

Recommended values of c_c, e and k given by the British Steel Piling Company Limited are shown in Figure 11.28 and Tables 11.6 and 11.7.

Type of hammer

The pile driving formulas are based on free falling drop hammers. If single-acting steam or compressed air

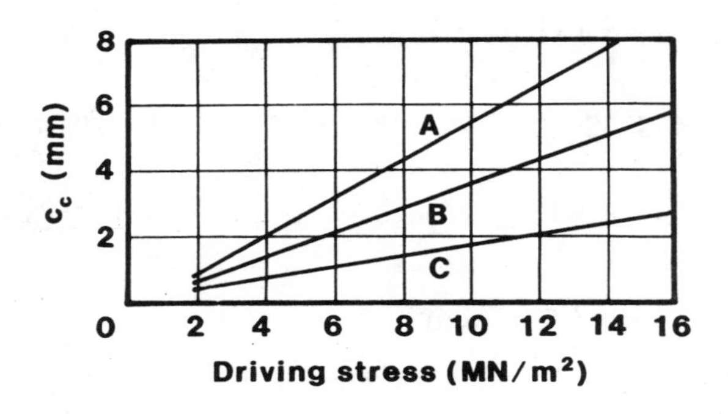

Figure 11.28 Values of c_c recommended by the British Steel Piling Company Limited:

A - 75mm packing under helmet with a concrete pile
B - helmet with dolly on a concrete or steel pile
C - 25mm pad only on a concrete pile.

TABLE 11.6
VALUES OF HAMMER COEFFICIENT k

Hammer	k
Drop hammer operated by a trigger release	1.0
Drop hammer operated by releasing the winch clutch and overhauling the rope.	0.8
Single-acting steam or compressed air hammer.	0.9

TABLE 11.7
VALUES OF COEFFICIENT OF RESTITUTION e

Type of pile	Head condition	Drop single-acting or diesel hammers	Double-acting hammers
Reinforced concrete	Helmet with composite plastic or greenheart dolly and packing on top of pile.	0.4	0.5
	Helmet with timber dolly (not greeheart), and packing on top of pile.	0.25	0.4
	Hammer direct on pile with pad only.	-	0.5
Steel	Driving cap with standard plastic or greenheart dolly.	0.5	0.5
	Driving cap with timber dolly (not greenheart)	0.3	0.3
	Hammer direct on pile.	-	0.5
Timber	Hammer direct on pile.	0.25	0.4

hammers are used, the weight W and drop H of the moving parts can be considered as for a drop hammer and losses due to piston friction and back pressure can be allowed for in the values of k. However, with double-acting hammers, the energy per blow must be obtained from manufacturers' specifications, and will depend on the speed of the hammer. This will replace the value of WH or kWH in the formulas. The rated energy per blow can be checked by alternately driving with a diesel or steam hammer and a drop hammer.

PILE LOAD TESTS

Pile bearing capacity formulas should not be expected to give more than a rough indication of the ultimate load capacity of a pile and, except where piles are driven to refusal, it is usual to load test at least one pile at each site. Special test piles may be driven ahead of the main construction program and tested to failure. As a result of these tests, the engineer may decide to modify the pile lengths required or reappraise the value of the pile working load. It is preferable to delay testing a pile for as long as possible after it has been driven to allow it to settle down. This is not so important with piles in coarse granular soils, where time-dependent effects are negligible, but in silts and silty sands the ultimate capacity of a pile may be much higher immediately after driving than after it has been installed for a month or so. In clays, the reverse is usually (but not always) true; the carrying capacity increasing with time, particularly in soft or sensitive clays.

Pile load tests may be carried out to give the load/settlement characteristics of a pile and to check its carrying capacity. The pile is usually tested to some multiple of working load (typically 1.5 or 2 times working load), or to failure.

Application of load

Two methods are commonly used:

a. the "kentledge" method, in which the pile is jacked against a heavy weight, or kentledge, supported above the pile on a framework of steel beams;
b. the "boot strap" method, in which the pile is jacked against a steel beam which is anchored to two or more tension piles.

Measurement of load and deflection

The load applied to the pile is usually measured by means of a proving ring or load cell.

Settlement of the pile can be measured by attaching a survey target to it and accurately levelling the target with reference to a fixed datum located at least 10m from the test set up. The system must have an accuracy of at least 1mm. Greater precision can be obtained using dial gauges attached to a beam that is supported by stanchions embedded in the ground at least 5m from the pile. The gauges bear on to glass plates set into the top of the pile. However, although the dial gauge system may appear to be much more accurate than direct levelling, results may be affected by movement of the ground around the stanchions supporting the beam that carries the dial gauges. It is better to use both systems of measurement, as a check.

The maintained load test

The load is applied in stages, typically in increments which represent about 25% of the working load, although initial increments may be larger and later increments smaller. Once each load increment has been applied, the load is kept constant until the rate of settlement of the pile is very low. British Standard Code of Practice CP 2004 specifies a maximum rate of 0.1mm in 20 minutes and ASTM specification D1143-57T specifies that the load be maintained for two hours or until the rate of settlement is not more than 0.305mm/hr (0.001ft/hr), whichever occurs first.

The pile may be unloaded completely at working load and then reloaded in one stage to the next load increment. Final unloading at the end of the test may be carried out in one or more stages.

Where the test is used to determine the ultimate bearing capacity of the pile, failure is usually defined as the load which causes settlement equal to 10% of the pile diameter, or at which the rate of settlement continues undiminished for a constant load.

Constant rate of penetration test

The pile is pushed into the ground at a constant rate

of penetration, and the force required is measured. As the pile penetrates the ground, the force required will steadily increase, usually reaching a maximum value, which is the ultimate bearing capacity of the pile. For end-bearing piles, a maximum will not usually be reached and the ultimate bearing capacity is taken as the force at a penetration equal to 10% of the pile diameter.

This test is generally considered to be more suitable for determining ultimate bearing capacity than the maintained load test, but the load/deflection characteristics are quite different from those of the maintained load test and cannot be used to predict settlement of the pile under working load conditions.

The force is usually applied by hydraulic jack, which should have a travel of at least 20% of the pile diameter, plus 75mm to allow for movement of the kentledge, or 25mm if tension piles are used.

For friction piles (in clay), a rate of penetration of 0.75mm/min is convenient. For end-bearing piles, a rate of 1.5mm/min is more suitable.

Interpretation of pile load test results using the Chin plot

The results of a pile load test are normally reported in the form of a plot of load on the pile head (Q) against settlement of pile head (Δ). In recent years increasing use has been made of an additional plot of settlement/load, Δ/Q, against settlement, Δ. This additional plot, termed the Chin plot after its originator, F.K. Chin, provides a basis for:

a. separating the pile's load capacity into its two components of shaft friction and end bearing;
b. estimating the ultimate load of a pile from the results of a routine pile load test even though the test is not taken anywhere near to failure; and
c. assessing the structural integrity of a pile in terms of possible damage or deviation from linearity during installation or on loading.

At first glance, it seems surprising that such a simple plot could provide so much additional information, but a brief explanation of Chin's hypothesis shows that it is founded on an ingenious and convincing theoretical approach.

Chin's hypothesis is based on two main assumptions:

1. the plot of load on pile head (Q) against settlement of pile head (Δ) gives a rectangular hyperbola possessing the equation

$$Q = \frac{\Delta}{m\Delta + c} \qquad (3)$$

   where m and c are constants defining the interaction between the pile and the soil (Figure 11.29(a)); and
2. on application of load, friction on the pile shaft is mobilised progressively from the ground surface downwards, with maximum shaft friction achieved before any end bearing is developed.

Re-writing equation (3) in the form

$$Q = \frac{1}{m + (c/\Delta)} \qquad (4)$$

shows that $Q \to 1/m$ as $\Delta \to$ a high value

This asymptotic behaviour is a fundamental property of a rectangular hyperbola and, for the Q against Δ relationship shown, leads to the conclusion that

$$\text{pile ultimate load } Q_u = 1/m.$$

So if the constant m can be determined for the hyperbola, the pile ultimate load is obtained.

To determine m, re-write equation (3) in the form

$$\Delta/Q = m\Delta + c \qquad (5)$$

This has the form $y = mx + c$, the equation of a straight line. A plot of (Δ/Q) against Δ should, therefore, yield a straight line of slope m, thus allowing determination of $Q_u (= 1/m)$. When the results of a pile load test are plotted in this manner, however, two straight lines AB and BC usually result (Figure 11.29(b)).

To explain this, it is necessary to examine Chin's second assumption that maximum shaft friction is achieved before end bearing is developed. The implication of this assumption is two separate phases of load capacity generation governed by two different pile/soil interaction mechanisms: firstly, the build-up of shaft friction caused only by elastic compression of the pile body; secondly, the build-up to ultimate load as end bearing is developed by penetration of the pile toe into the soil. These two interaction mechanisms possess different m and c values and, hence, two different straight lines on the plot of Δ/Q against Δ. This explanation leads to the conclusion that, in Figure 11.29(b), line AB represents the build-up of shaft friction with maximum shaft friction being $F_{max}=1/m_{AB}$ and line BC represents the build-up to ultimate load with $Q_u=1/m_{BC}$.

A routine pile load test, taken typically to 1.5 or 2.0 times the allocated working load, invariably reaches a pile head settlement in excess of that required to draw line AB with sufficient additional points to define the slope of line BC. An estimate can therefore be made of the ultimate load of the pile even though the test may not have been taken anywhere near to failure. A pile that has suffered damage or has deviated from its intended linear shape during installation can be identified by its non-standard Chin plot, which will often reveal imperfections not highlighted by the usual Q vs. Δ plot. A typical example is shown in Figure 11.29(b).

The accuracy of the Chin predictions is dependent upon the validity of the initial assumptions made in

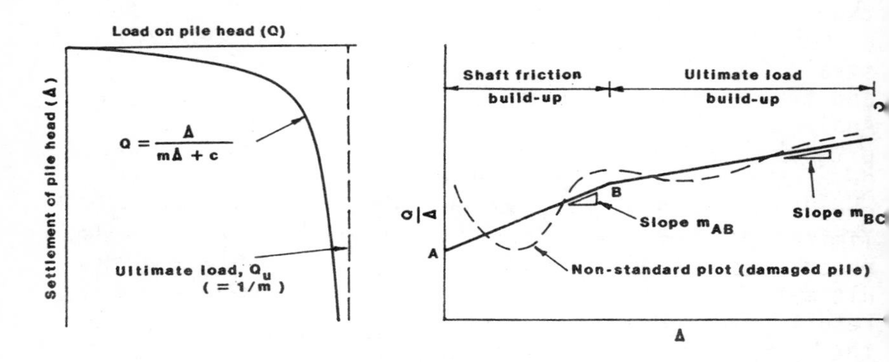

Figure 11.29 Chin's hypothesis: (a) assumed load/ displacement relationship; (b) the Chin plot.

formulating the hypothesis. Most piles do exhibit a hyperbolic Q against Δ plot under test loading but few actually reach the limiting value shown in Figure 11.29(a). The criterion adopted for estimation of ultimate load is usually that of the load corresponding to a pile head settlement of 1/10th of the pile diameter, and the divergence of the theoretical hyperbolic curve and the actual curve is at its greatest at such high values of settlement. The second assumption of progressive mobilisation of shaft friction to its maximum value before any development of end bearing is also questionable. Evidence of build-up of some end bearing from the beginning of test loading is contained in published literature and is commonplace in most model-scale research programmes. These and other limitations ensure that the hypothesis cannot be completely correct. Present evidence suggests that the hypothesis significantly overestimates the value of maximum shaft friction in many cases but provides a more accurate assessment of ultimate load (giving, on average, an overestimate of 30%). The presence of damaged piles is consistently revealed.

Dynamic load testing of piles

The last few years have seen a steady increase in the use of dynamic pile testing. The pile under test is struck at the top, typically using a heavy hammer or the pile driving hammer, but with a drop of only about a metre. The resulting shock wave is picked up by strain transducers, attached near the top of the pile, and transmitted to a recording device, in digital form. Analysis is carried out using standard computer programs, either on site or at the home base. Originally, this type of testing was used only as a check on the integrity of the pile itself, and such limited systems are still in common use. More recently-developed systems are also able to predict the ultimate bearing capacity of the pile. However, when results are compared with static load tests, errors of the order of ±30% are not uncommon, and both over-estimates and under-estimates may occur on the same site. Because of this, it is still usual to supplement dynamic testing with traditional static tests, but to use the relative cheapness of the dynamic method to increase the number of piles tested.

Chapter 12

EARTH RETAINING WALLS AND SLOPES

TYPES OF WALL

A brief review of earth retaining walls is given in Chapter 2. Viewed in terms of stability analysis, walls may be divided broadly into rigid walls and flexible walls. The analysis for rigid walls depends on simple statics: earth pressure forces behind the wall, which tend to cause it to slide or topple over, are resisted by stabilising forces, which result from the weight of the wall and friction along its base. Analysis consists mainly of equating horizontal thrusts to check for resistance to sliding; and equating moments about the toe to check for resistance to overturning. Flexible walls are embedded in the ground and rely for stability on movement of the wall and of the soil masses behind and in front of them. The weight of the wall itself plays no part in the stability. As the wall moves forwards, under the influence of the pressure of the ground behind it, earth pressure behind the wall decreases until it reaches a minimum value. At the same time pressure on the front of the wall increases to a maximum value as the earth at the front is pushed forwards. The stability of the wall is calculated by comparing the minimum thrust from the backfill with the maximum resistance offered by the ground in front of the wall and by any anchorages which may be installed.

EARTH PRESSURES ON WALLS

The first step when calculating the stability of earth retaining structures is to estimate the thrust exerted on the structure by the earth behind it. Two concepts are useful when considering how the soil acts on the back of the wall. The first concept assumes that the soil acts like a (sort-of) liquid, in which pressures within the mass are related to depth and density, as illustrated in Figure 12.1(a). The difference between

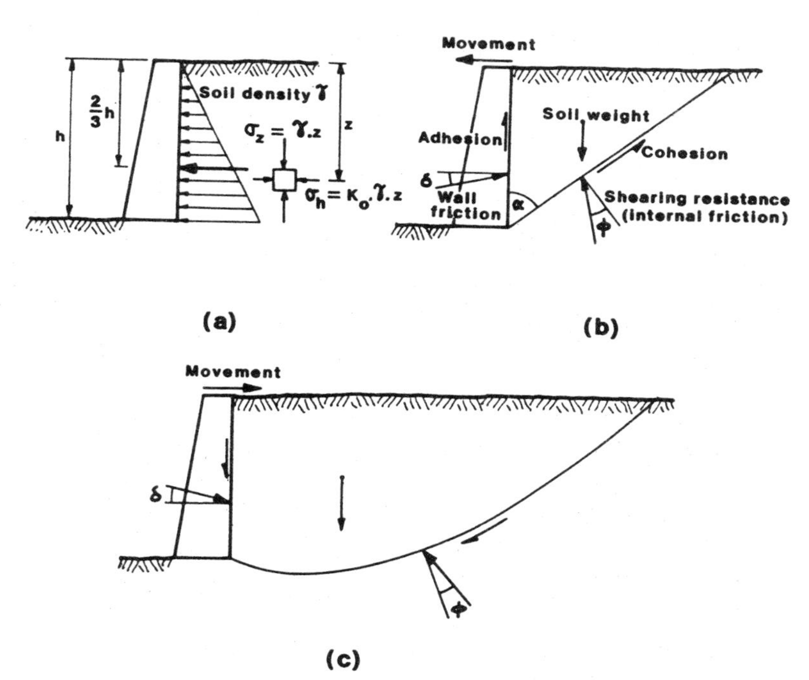

Figure 12.1 Concepts of soil pressure behind retaining walls: (a) coefficients of earth pressure; (b) a wedge of soil sliding on a failure plane; (c) soil sliding over a curved failure surface.

the soil mass and a true liquid is that, because the soil has shear strength, horizontal pressures need not be the same as vertical pressures. It is assumed that horizontal pressures are some proportion K of vertical pressures. In an undisturbed soil mass, $K = K_o$, the coefficient of earth pressure at rest. The value of K_o depends on the soil type and its geological history. Values of K_o of 0.6 to 0.8 are typical. Thus, the pressure on the back of the wall shown in the figure is

$K_o.\gamma.z$ at depth z. The total force P on the wall will be $P=\frac{1}{2}K_o.\gamma.h^2$, acting $^2/_3$ of the way down the wall. What happens to the earth pressure if the wall moves forwards can best be understood by considering the second concept: to imagine a block of soil behind the wall which slides with it, as shown in Figure 12.1(b). The friction and cohesion forces that develop on the base of the block as it slides downwards act upwards and backwards, away from the wall. This reduces the value of thrust P due to the soil. The thrust can be calculated by trying various values of wedge angle α to find the worst possible case. This turns out to be when $\alpha = 45^0+\frac{1}{2}\phi$. This approach is known as the Coulomb method of analysis. Using this method, friction and adhesion at the back of the wall can also be taken into account. If the wall were pushed back into the soil instead of being allowed to move forwards then the cohesion and friction forces on the wedge would be reversed, increasing the thrust on the wall. In fact, the critical shape for this condition turns out to be a spiral, as shown in Figure 12.1(c), not a wedge.

The situations described above are known as the active condition, when the soil pushes the wall forwards; and the passive condition, when the soil resists backward movement of the wall. For these two limiting conditions the concept of a constant ratio between vertical and horizontal stresses holds good only for cohesionless soil. However, long term drained conditions are usually the most critical for retaining walls and since even clays show little or no cohesive strength under drained conditions, the concept is a useful one for all soil types. The coefficient of earth pressure in the active condition, when horizontal thrusts are at their lowest, is known as the coefficient of active earth pressure, K_a. The maximum value of horizontal to vertical pressures, developed in the passive condition, is known as the coefficient of passive pressure, K_p. For retaining wall design, values of active and passive pressure coefficients are normally taken from curves by Caquot and Kerisel, who assumed a logarithmic spiral failure surface in their

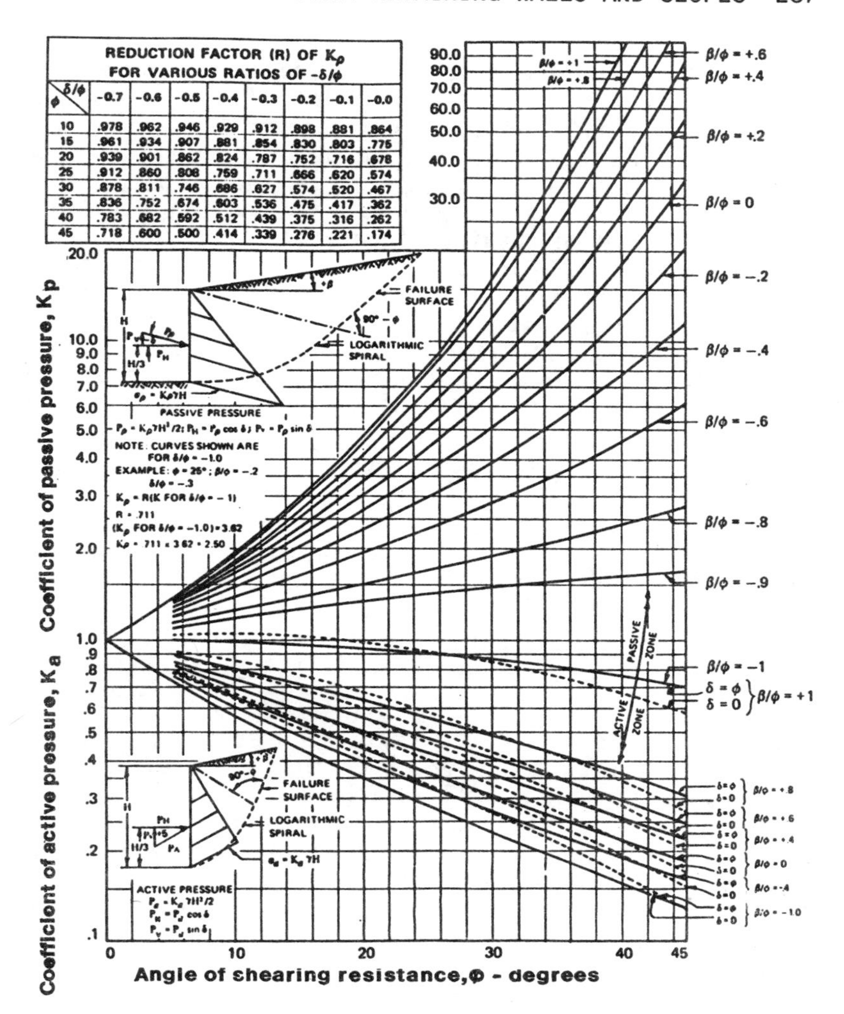

REDUCTION FACTOR (R) OF K_p FOR VARIOUS RATIOS OF $-\delta/\phi$								
ϕ \ δ/ϕ	-0.7	-0.6	-0.5	-0.4	-0.3	-0.2	-0.1	-0.0
10	.978	.962	.946	.929	.912	.898	.881	.864
15	.961	.934	.907	.881	.854	.830	.803	.775
20	.939	.901	.862	.824	.787	.752	.716	.678
25	.912	.860	.808	.759	.711	.666	.620	.574
30	.878	.811	.746	.686	.627	.574	.520	.467
35	.836	.752	.674	.603	.536	.475	.417	.362
40	.783	.682	.592	.512	.439	.375	.316	.262
45	.718	.600	.500	.414	.339	.276	.221	.174

Figure 12.2 Active and passive earth pressure coefficients for granular soil. (After Caquot and Kerisel).

calculations. Their curves for a vertical wall and sloping backfill are given in Figure 12.2. Note that the K_p values given are for a wall friction angle δ equal to the angle of shearing rsistance (i.e. $\delta/\phi=1$). For lesser angles of δ the value of K_p obtained from the graph must be multiplied by a reduction factor, obtained from the inset table. For design purposes, values of δ may be obtained from Table 12.1.

An alternative approach to obtain active and passive pressure coefficients is to consider failure conditions in cohesionless soils. Figure 12.3 shows the Mohr-Coulomb failure envelope for a cohesionless soil and a Mohr circle representing a state of stress at failure. From the geometry of the diagram.

$$\sin\phi = \frac{BC}{OC} = \frac{\frac{1}{2}(\sigma_1 - \sigma_3)}{\frac{1}{2}(\sigma_1 + \sigma_3)}$$

rearranging gives:

$$\frac{\sigma_3}{\sigma_1} = \frac{1 - \sin\phi}{1 + \sin\phi} \quad \text{or} \quad \frac{\sigma_1}{\sigma_3} = \frac{1 + \sin\phi}{1 - \sin\phi}$$

In all conditions the vertical pressure σ_z is that due to the overburden, $\gamma.z$ at depth z. If the soil is behind a retaining wall which is allowed to move forwards, horizontal stress σ_h will be reduced until failure occurs within the soil mass, when σ_h will reach a minimum value, σ_a, the active pressure. Thus, the vertical pressure, σ_z, will become the maximum principal stress σ_1 and the horizontal pressure, σ_a, will be the minimum principal stress, σ_3. We can then define the coefficient of active pressure as:

$$K_a = \frac{\sigma_a}{\sigma_z} = \frac{\sigma_3}{\sigma_1} = \frac{1 - \sin\phi}{1 + \sin\phi} = \tan^2(45^0 - \frac{1}{2}\phi)$$

If the wall is pushed towards the soil then horizontal pressure will increase until failure occurs by the soil being pushed outwards and backwards, when σ_h will reach a maximum value σ_p, the passive earth pressure. The horizontal pressure will now be greater

TABLE 12.1
TYPICAL VALUES OF FRICTION AND ADHESION AT INTERFACES

Materials	Friction factor, $\tan\delta$	Adhesion factor, c_a (kN/m^2)
Mass concrete or masonry against rocks and soils:		
Clean sound rock	0.7	
Clean gravel, gravel-sand mixtures, coarse sand	0.55-0.60	
Clean fine to medium sand, silty medium to coarse sand, silty or clayey gravel	0.45-0.55	
Clean fine sand, silty or clayey fine to medium sand	0.35-0.45	
Fine sandy silt, non-plastic silt	0.30-0.35	
Very stiff and hard residual or overconsolidated clay	0.40-0.50	
Stiff clay and silty clay	0.30-0.35	
Steel sheet piles against soils:		
Clean gravel, gravel-sand mixtures, well-graded rock fill	0.40	
Clean sand, silty sand-gravel mixtures, single size hard rock fill	0.30	
Silty sand, gravel or sand mixed with silt or clay	0.25	
Fine sandy silty, non-plastic silt	0.20	
Soft clay and clayey silt		5-30
Stiff and hard clay and clayey silt		30-60
Formed concrete or concrete sheet piling against soils:		
Clean gravel, gravel-sand mixtures, well-graded rock fill	0.40-0.50	
Clean sand, silty sand-gravel mixtures, single size hard rock fill	0.30-0.40	
Silty sand, gravel or sand mixed with silt or clay	0.30	
Fine sandy silt, non-plastic silt	0.25	
Soft clay and clayey silt		10-35
Stiff and hard clay and clayey silt		35-60
Various structural materials:		
Masonry on masonry, igneous and metamorphic rocks:		
-dressed soft rock on dressed soft rock	0.70	
-dressed hard rock on dressed soft rock	0.65	
-dressed hard rock on dressed hard rock	0.55	
Masonry on wood (cross grain)	0.50	
Steel on steel at sheet pile interlocks	0.30	

Note: the numbers are ultimate values and require sufficient movement for failure to occur. Where friction factor only is shown, the effect of adhesion is included in the friction factor. For data on adhesion on bearing piles, see Chapter 11.

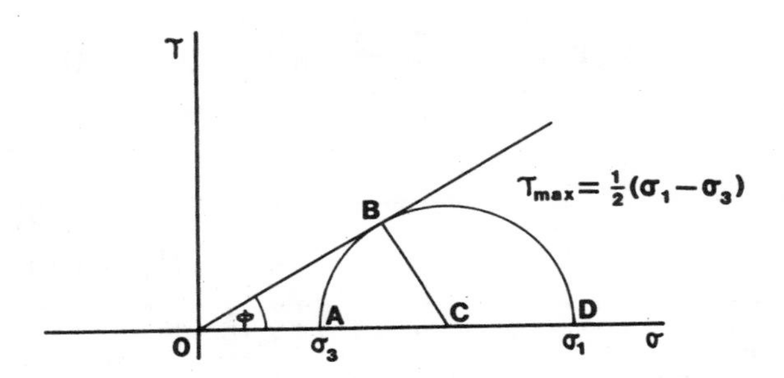

Figure 12.3 Ratio of maximum and minimum principal stress for a granular soil.

than the vertical pressure, so in this case $\sigma_p = \sigma_1$ and $\sigma_z = \sigma_3$. The coefficient of passive pressure can be defined as:

$$K_p = \frac{\sigma_p}{\sigma_z} = \frac{\sigma_1}{\sigma_3} = \frac{1 + \sin\phi}{1 - \sin\phi} = \tan^2(45^0 + \tfrac{1}{2}\phi)$$

This approach is known as Rankine's method. Because it considers only stresses within the soil mass it cannot take into account wall friction. The resulting errors are not significant for K_a values but K_p values are grossly underestimated, leading to over-conservative designs. For instance, for a soil with a horizontal surface and a shearing resistance $\phi = 34^0$ and a wall friction $\delta = 17^0$ (giving $\delta/\phi = 0.5$), Figure 12.2 gives

$$K_p = 9.5 \times 0.674 = 6.4$$

Rankine's method gives

$$K_p = \tan^2(45^0 + \tfrac{1}{2}\times 34^0) = 1.9$$

The main advantage of this approach is that it can be modified to give a relatively simple method of calculating coefficients of active and passive pressure for soils possessing cohesion as well as internal friction. Figure 12.4 shows a Mohr-Coulomb failure envelope for the general case of a c-ϕ soil. By analogy with Figure 12.3 and the expressions developed for a cohesionless soil it can be seen that

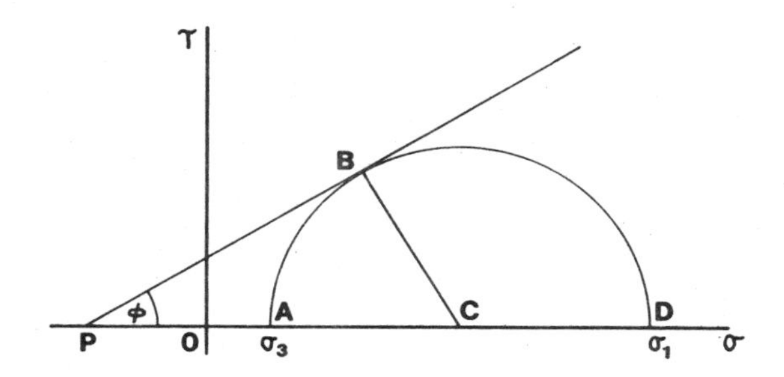

Figure 12.4 The relationship between maximum and minimum principal stresses for a c - φ soil: Bell's solution.

$$\frac{PA}{PD} = \frac{c.\cot\phi + \sigma_3}{c.\cot\phi + \sigma_1} = \frac{1 - \sin\phi}{1 + \sin\phi} = \tan^2(45^0 - \tfrac{1}{2}\phi)$$

For the active case, as discussed previously, $\sigma_1 = \sigma_z$ and $\sigma_3 = \sigma_a$. Making these substitutions and putting $\tan^2(45^0 - \tfrac{1}{2}\phi) = K_a$ gives, after some rearrangement,

$$\sigma_a = K_a\sigma_z + (K_a - 1).c.\cot\phi$$

But $$K_a - 1 = \frac{1 - \sin\phi}{1 + \sin\phi} - 1 = -\frac{2\sin\phi}{1 + \sin\phi}$$

So $$(K_a - 1)\cot\phi = \frac{-2\sin\phi}{1 + \sin\phi}.\frac{\cos\phi}{\sin\phi} = \frac{2\cos\phi}{1 + \sin\phi}$$

$$= -2\tan(45^0 + \tfrac{1}{2}\phi)$$

$$= -2\sqrt{K_a}$$

Therefore

$$\sigma_a = K_a\sigma_z - 2c\sqrt{K_a}$$

For the passive case, $\sigma_1 = \sigma_p$ and $\sigma_3 = \sigma_2$. By a similar process to that used for the active case, and putting $\tan^2(45^0 + \tfrac{1}{2}\phi) = K_p$, we obtain

$$\sigma_p = K_p\sigma_z + 2c\sqrt{K_p}$$

This method of estimating earth pressures for c-ϕ soils is known as Bell's solution. It can be seen that the formulas for active and passive pressures contain the constant term $2c\sqrt{K_a}$ or $2c\sqrt{K_p}$ so that lateral pressures are no longer simple ratios of vertical pressure. Values K_a and K_p are therefore not strictly coefficients of earth pressure as originally defined. Sometimes different symbols are used to emphasise this.

The formula for active pressure implies that active pressures will be negative near the surface, where $K_a\sigma_z < 2c\sqrt{K_a}$. However, it is usually assumed that tension cannot exist between the wall and the soil behind it, and active pressures are assumed to be zero down to a critical depth z_c which can be found by putting

$$\sigma_z = \gamma.z_c \text{ and } \sigma_a = 0, \text{ so that } K_a.\gamma.z_c = 2c\sqrt{K_a}$$

giving

$$z_c = \frac{2c}{\gamma\sqrt{K_a}} = \frac{2c}{\gamma.\tan(45^0 - \frac{1}{2}\phi)} = \frac{2.c.\tan(45^0+\frac{1}{2}\phi)}{\gamma}$$

For short term analysis in clays, the undrained shear strength is used and usually $\phi = 0$. This gives $K_a = K_p = \tan 45^0 = 1$. Then

$$\sigma_a = \sigma_z - 2c$$

$$\sigma_p = \sigma_z + 2c$$

For complex situations involving c-ϕ soils, more accurate predictions of stability can be achieved by analysing the stability of soil wedges behind the wall. Established procedures exist for this approach but calculations are complex and are avoided whenever possible.

In calculating values of σ_z for active and passive pressure calculations, effective stresses are used as indicated in Figure 12.5, unless short term conditions are analysed, when total stress analysis is used.

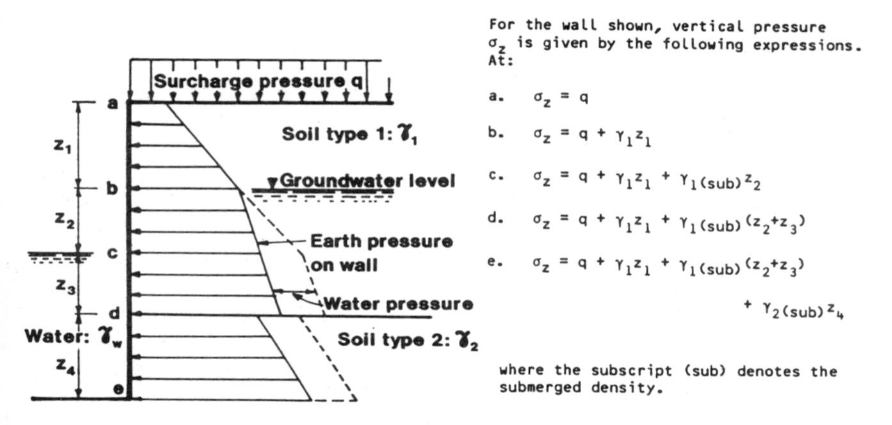

Figure 12.5 Examples of the calculation of vertical earth pressure for various conditions.

RIGID RETAINING WALLS

Earth pressures

Rigid retaining walls are normally built up to about 6m high. The amount of forward movement or tilting of the wall required for active conditions to develop may be unacceptably high for a rigid wall. Therefore, in order to avoid excessive movement, designs often assume higher values of lateral earth pressure than would occur in active conditions. Design values for earth pressures are often based on values suggested by Terzaghi and Peck, given in Figure 12.6. These are based on the assumption that the backfill is well drained.

Stability calculations

The forces acting on a retaining wall are shown in Figure 12.7. The wall can fail by either overturning or sliding but overturning is usually the critical mode of failure.

When considering overturning, the contribution of the soil in front of the wall is usually ignored. The first step is to find the point of action of resultant

DESIGN LOADS FOR LOW RETAINING WALLS

Design loads for walls less than about 6m high may be obtained from the charts below which are based on values suggested by Terzaghi and Peck. Soil types referred to are:-

1. Clean sand and gravel: GW, GP, SW, SP.
2. Silty sand and gravel: GM, GM-GP, SM, SM-SP.
3. Stiff natural deposits of silts and clays, silty fine sands, clayey sands and gravels: CL, ML, CH, MH, SM, SC, GC.
4. Very soft to soft clay, silty clay, organic silt and clay: CL, ML, OL, CH, MH, OH.
5. Medium stiff clay, deposited in chunks and protected from infiltration: CL, CH. For this material, H is reduced by 1.25m and the resultant acts at a height $\frac{1}{3}$(H - 1.25)m above the base.

The circled numbers in the graphs indicate soil types.

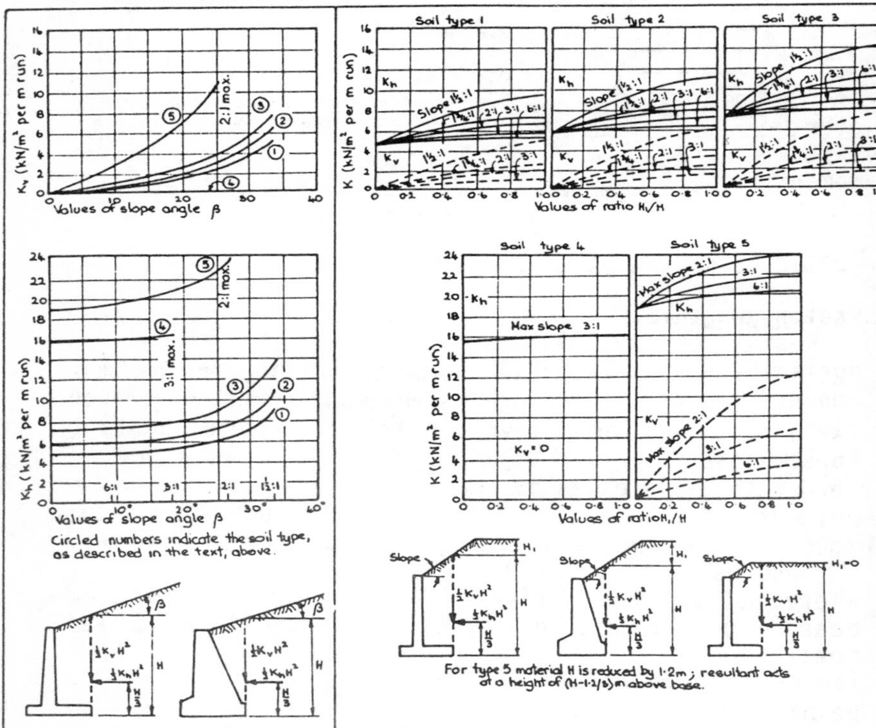

Figure 12.6 Suggested design values for earth pressures behind low rigid retaining walls.

R. This is done by taking moments about the toe:

$$d = \frac{W.a + P_v.e - P_h.b}{W + P_v}$$

where $(W + P_v)$ is the vertical reaction of the ground beneath the wall.

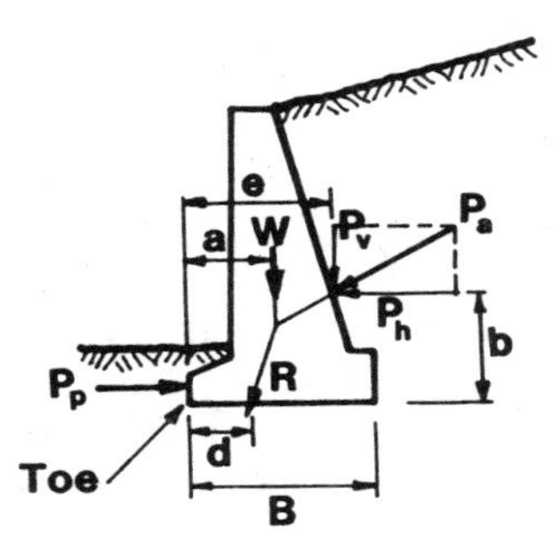

Figure 12.7 Forces acting on a rigid retaining wall.

If the point of action is within the middle third, overturning may be ignored. Otherwise the factor of safety F against overturning can be obtained by taking moments about the toe:

$$F = \frac{W.a}{P_h.b - P_v.e}$$

Factor of safety F should normally be at least 1.5.

Resistance to sliding is checked by comparing horizontal thrust from the backfill with the maximum amount of friction and adhesion that can be developed along the base. The factor of safety against sliding is thus:

$$F = \frac{(W + P_v)\tan\delta + c_a.B + P_p}{P_h}$$

where c_a and δ are the adhesion and friction along the base of the wall. It is usual to ignore the contribution of the soil in front of the wall (so $P_p = 0$), especially if there is a possibility of it being washed away or dug up, and to accept a minimum factor of safety of 1.5. If thrust P_p is taken into account, the minimum acceptable factor of safety is usually 2.

Choice of parameters

Weight W includes the weight of the wall and, in cantilever and counterfort walls, the weight of the soil above the footing, behind the wall (see Figure 2.10(a)). Values of P_a and its components P_v and P_h are obtained either from the analysis of active conditions behind the wall (using the curves given in Figure 12.2 or Bell's solution) or, for low walls, from the values given in Figure 12.6. Wall friction has little effect on values and is usually ignored. If the wall is on a soft foundation it may settle relative to the backfill causing a reversal of wall friction. In this case, values obtained from Figure 12.6 should be increased by 50%. Where passive thrust P_p is taken into account it is usual to ignore the effect of wall friction when calculating values of P_p. Values of adhesion c_a and friction angle δ along the base of the wall may be obtained from Table 12.1.

Water pressures

Water pressure must be taken into account below the standing water table. Values of active and passive pressure are reduced by water pressure because only the submerged density is relevant for overburden pressure calculations but the thrust on the wall due to the water itself must be taken into account. The overall effect is to considerably reduce stability and it is usually more economical to provide drainage, with weep holes through the wall. If earth pressures are obtained from Figure 12.6, drainage must be provided.

Settlement and bearing capacity considerations

It can be seen from Figure 12.7 that the resultant thrust on the base of the wall tends to act towards the front of the wall. This results in differential settlement of the base and forward tilting of the wall. Bearing pressures across the base can be estimated using the technique described in Chapter 11 for foundations with an eccentric loading, where the total downward thrust Q on the base is W + P_v acting at a distance d from the toe. Settlement and differential settlement can then be estimated from consolidation theory or elastic theory, in the usual way. If the consequent tilt of the wall is likely to exceed about $\frac{1}{2}^{0}$ then the base should be reproportioned to keep the resultant nearer the centre.

The maximum bearing pressure obtained from these calculations must not exceed the allowable bearing pressure for the soil.

BRACED EXCAVATIONS

Bracing generally consists of vertical sheet piling supported by a series of struts and walings. The construction sequence is usually

(a) steel sheet piles are driven into the ground
(b) ground is excavated from inside
(c) walings and struts are installed and tightened as excavation proceeds.

Because of the method of construction, which allows relaxation of the sides before the struts are fixed in position, pressures on the sheet piling are usually lower than predicted values using traditional earth pressure theories. The usual design procedure follows the method proposed by Terzaghi, using rules similar to those given below (see Figure 12.8).

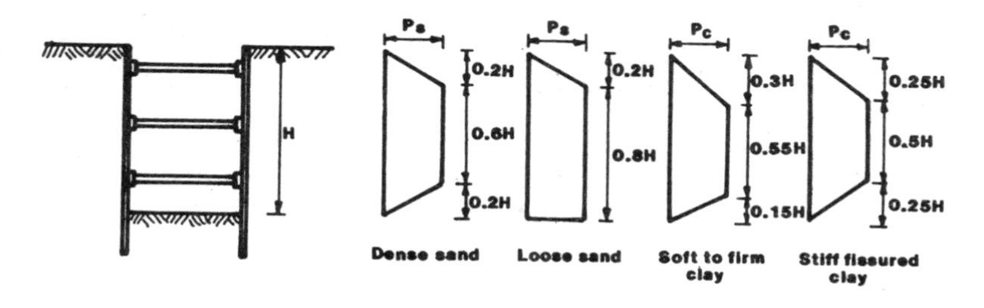

Figure 12.8 Design pressures for braced excavations.

In sand, pressure $p_s = 0.8K_a\gamma.H.\cos\delta$, where

K_a is the active pressure determined from Figure 12.2
γ is the average soil density
H is the depth of excavation
δ is the angle of wall friction, from Table 12.1

In clay, pressure $p_c = \gamma H - 4c_u$, where
c_u is the undrained shear strength.
In stiff fissured clays $p'_c = 0.4.\gamma.H.$

If movement can be reduced to a minimum and construction time is short, this can be reduced to

$$p'_c = 0.2.\gamma.H.$$

The submerged density is used below the water table and water pressures are added to the calculated earth pressures.

In clays, creep effects cause a redistribution of stresses with time. From model tests carried out by Kirkdam, it was concluded that, for long-term design in clays, classical earth pressure theories should be used, based on effective stress parameters.

CANTILEVER RETAINING WALLS

Sheet piling is driven into the ground to sufficient depth for it to become fixed as a cantilever, resisting pressure from the backfill side. The piling will be pushed forward by the backfill and will tend to rotate about a point near the base, as shown in Figure 12.9. This will result in forces being developed as illustrated in Figure 12.10.

Pivot point O can be obtained by trial and error, balancing moments. However, it is usual to assume that the passive force on the base of the piling can be represented by a single horizontal thrust R, resulting in a force diagram of the type shown in Figure 12.11. This greatly simplifies the design calculations without introducing serious errors.

Stability calculations

The first stage in the analysis is to determine active and passive pressures and water pressures either side of the wall. Figure 12.11 illustrates a very simple case, with only one soil type and no water pressures. The minimum driving depth d can be obtained by taking moments about the base of the piling. In the example shown:

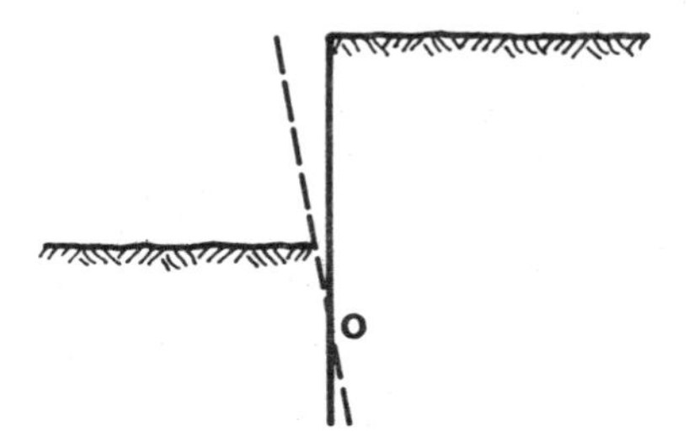

Figure 12.9 Movement of a cantilever retaining wall.

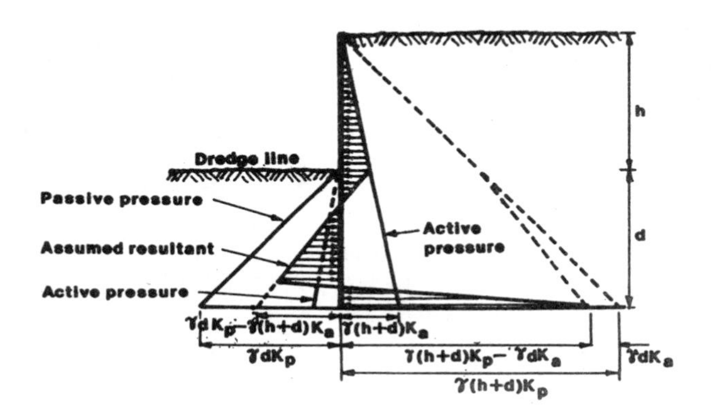

Figure 12.10 Conventional assumed earth pressure distribution on a cantilever retaining wall.

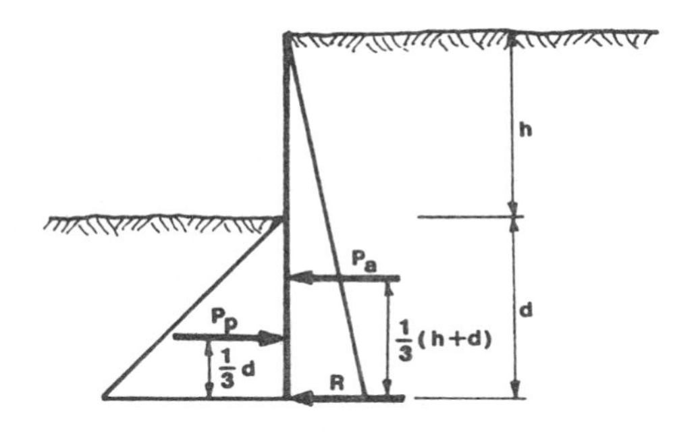

Figure 12.11 Simplified force diagram for a cantilever retaining wall.

$$\tfrac{1}{3}P_p \cdot d = \tfrac{1}{3}P_a(h + d)$$

Where appropriate, water pressures must be included in the calculations.

The wall is driven at least 20% beyond the calculated depth d, to give a factor of safety and allow the passive thrust R to be developed. Alternatively, a factor of safety is probably better introduced by dividing the coefficient of passive resistance by a factor (usually 1.5). This will result in a higher calculated depth d.

Where clay backfill is used the procedure is basically the same but wall pressures are calculated using Bell's solution. Stability is checked for both short and long term conditions, using undrained and drained strength parameters, respectively.

Shear forces and bending moments on the piling

The values of shear force and bending moment at any point along the sheet piling can be calculated by summing all the horizontal forces and moments, respectively, above or below that point, in the usual method of structural analysis.

ANCHORED RETAINING WALLS

Like cantilever walls, anchored walls are supported by the passive pressure of the ground in front of the wall, but in addition, anchored rods support the wall near the top. This reduces the length of embedment and the thickness of section required. Anchored sheet pile walls are suitable for heights up to about 10-12m, depending on soil conditions. For greater heights, high strength or reinforced sheet piling, extra tiers of tie rods, or relieving platforms may be required.

The stability of anchored walls and the stresses developed in them depend on the interaction between wall, anchors and soil. In general, a greater depth of penetration reduces flexural stresses. Because of the redundant nature of the problem, a rigorous analysis is not possible and a number of design methods have been developed, which fall broadly into "fixed earth" and "free earth" support theories. A simplified form of each, suitable for most design purposes, is given below.

Free earth support method

It is assumed that the depth of penetration of the wall is insufficient to enable the toe to be rigidly fixed, as it is with cantilever walls, and that the toe is pushed forward until full passive resistance is developed in front of it. The wall itself is assumed to be inflexible so that it rotates about the anchor point, which moves forward sufficiently for active pressures to be developed behind the wall. The forces on the wall for a simple case are illustrated in Figure 12.12. For active and passive pressures, the resultant can be calculated in the usual way. In the case illustrated,

$$P_a = \tfrac{1}{2}K_a(h + d)^2$$

acting $^2/_3$ of the way down AE, and

$$P_p = \tfrac{1}{2}K_p . d^2$$

acting $^2/_3$ of the way down CE. Out-of-balance water pressures must also be included in the calculation, if present.

The required depth d, for stability, can then be obtained by taking moments about B, the anchor point,

$$P_p(h_1 + {}^2/_3 d) = P_a[{}^2/_3(h + d) - (h - h_1)]$$

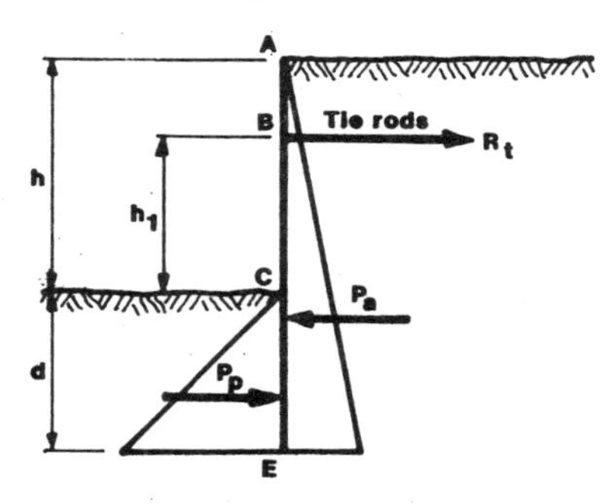

Figure 12.12 Simplified force diagram for an anchored retaining wall using the "free earth support" theory.

The value of d is obtained by trial and error.

The force R_T on the tie rods is obtained by equating horizontal forces once d has been determined. This will, of course, be the force per unit length of wall; the actual force on each tie rod will depend on the rod spacing. A factor of safety can be introduced either by driving the piling 40% greater than the calculated value of d or, preferably, by dividing the coefficient of passive pressure by a factor (usually 2). Shear forces and bending moments in the piling can be obtained in the usual way by summing forces and moments, respectively, above or below the point being considered. The position of the maximum bending moment occurs where shear forces are zero.

Actual pressures on the wall, and bending moments induced in it, are lower than those calculated by this method because of the flexibility of the wall. Bending moments calculated from this method are often multiplied by a reduction factor, based on work carried out by Rowe. The amount of reduction actually achieved depends on the flexibility of the wall and on the amount of arching in the material behind it. Such factors as the method of backfilling used and the presence of vibration can have a profound effect on the amount of arching. Because of this, some engineers are reluctant to apply reduction factors.

Fixed earth support (equivalent beam) method

The buried portion of the wall is assumed to be fixed, as with a cantilever wall. The full procedure requires the deflected shape to be calculated for a number of assumed values of buried depth d: the correct depth is found when the deflected shape of the wall passes through anchor point B. This method is very tedious and rarely used in practice. Instead, it is assumed that no moments can be transmitted through the point of contraflexure at D, shown in Figure 12.13, which acts as a hinge. The wall can then be considered as two simply supported beams, AD and DE, joined by a hinge. This is known as the equivalent beam method.

The value of depth x to the point of contraflexure (point D) is first assumed. Suitable values, obtained by Terzaghi, using a graphical integration method, are given in Figure 12.14; alternatively, a value of 0.1h

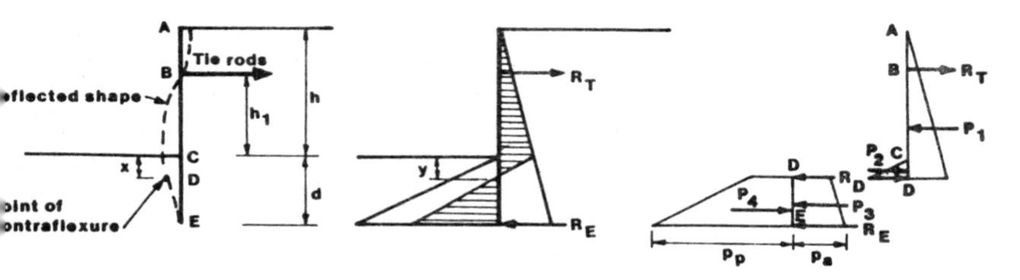

Figure 12.13 Forces on an anchored retaining wall using the "fixed earth support" (equivalent beam method).

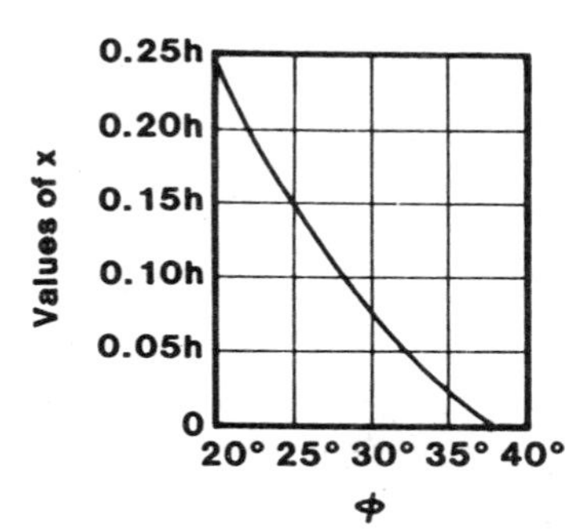

Figure 12.14 Values of depth x to point of contraflexure.

may be chosen. The upper portion (AD) can then be treated as a beam, simply supported at B and D. The lower portion can be treated in a similar manner, simply supported at D and E, with a force R_E to represent the fixity provided by the soil support at E.

Active and passive pressures (and out-of-balance water pressures, where present) are calculated. For the simple case shown in Figure 12.13, forces on the upper beam AD are

$$P_1 = \tfrac{1}{2}K_a \cdot \gamma \cdot (h + x)^2$$
$$P_2 = \tfrac{1}{2}K_p \cdot \gamma \cdot x^2$$

Reaction R_D at the hinge can be found by taking moments about point B:

$$R_D(h_1+x) = P_1[\tfrac{2}{3}(h+x) - (h-h_1)] - P_2(h_1+\tfrac{2}{3}x).$$

The force R_T on the tie rods (per unit length of wall) can be obtained by equating horizontal forces or by taking moments about D. Loads on beam DE consist of reaction R_D, active and passive soil pressures P_3 and P_4 (and any out-of-balance pore water pressures) and unknown reaction R_E at E. Forces P_3 and P_4 can be calculated, in terms of depth d, and moments of P_3, P_4 and R_D about point E can be equated to zero to solve for the value of d. Alternatively, d, can be found by trial and error.

For a wall with no out of balance pore water pressures, depth y to the point of zero net pressure is:

$$y = \frac{qK_a}{\gamma(K_p - K_a)} \quad \text{(see Figure 12.15),}$$

where q is the effective overburden pressure behind the wall at the dredge line. In this case, the required depth of penetration is approximately:

$$d = y + \sqrt{\frac{6R_D}{\gamma(K_p - K_a)}}$$

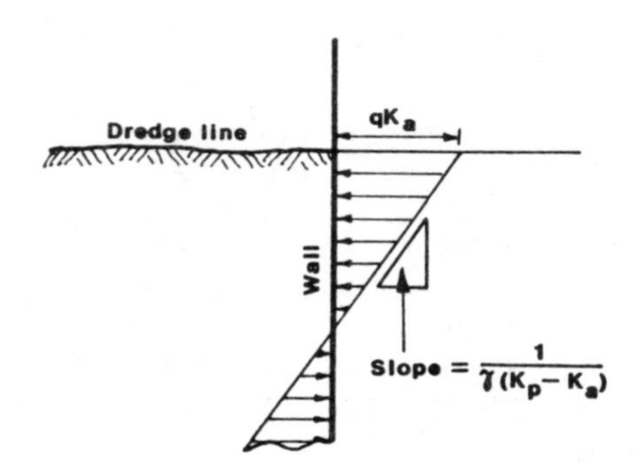

Figure 12.15 Point of zero net pressure on an anchored wall.

Moments and shear stresses can be obtained at any point on the section by summing moments and horizontal thrusts, respectively, above or below the point being considered, remembering that no moments are transmitted across point D.

Choice of methods

The fixed earth support, equivalent beam, method can be used only for sands. Use of this method where the base of the wall is embedded in medium to very dense sands gives a lower predicted bending moment than the "free earth" theory, resulting in a lighter section and more economical design. In these conditions, the assumption that the base of the wall is held rigidly is reasonable and the resulting design should be adequate.

The free earth support method may be used for all soil types and should result in an adequate design. This method is used where the wall is embedded in clay, when the equivalent beam method cannot be used, and probably gives more realistic values where the wall is embedded in loose sands which allow rotation of the base.

Design of anchorages and tie rods

The rods which restrain the upper part of the wall are normally tied back to an anchor wall or separate anchor blocks. The anchor must be located beyond the active wedge so that resistance can be fully developed, as shown in Figure 12.16.

For the anchor wall to the right of line ab and $h_1 > \frac{1}{2}H$

$$P_p = \tfrac{1}{2}K_p.\gamma.h^2$$
$$P_a = \tfrac{1}{2}K_a.\gamma.h^2$$
$$R_T = P_p - P_a = \tfrac{1}{2}\gamma.h^2(K_p - K_a)$$

where K_a and K_p are obtained in the usual way. For a concrete anchor wall, $\delta/\phi = 0.5$ is typically used when obtaining K_p.

For the anchor wall to the left of ab, as illustrated in the subsidiary sketch,

$$P_p = \tfrac{1}{2}K_p.\gamma.h^2 - (\tfrac{1}{2}K_p.\gamma.h_2^2 - \tfrac{1}{2}K_a.\gamma h_2^2)$$
$$P_a = \tfrac{1}{2}K_a.\gamma.h^2$$
$$R_T = \tfrac{1}{2}\gamma.(h^2 - h_2^2)(K_p - K_a)$$

For $h_1 < \frac{1}{2}h$, the wall is treated as a strip footing of width h, and surcharge load $(h - \frac{1}{2}h_1)$, using Terzaghi's bearing capacity formula.

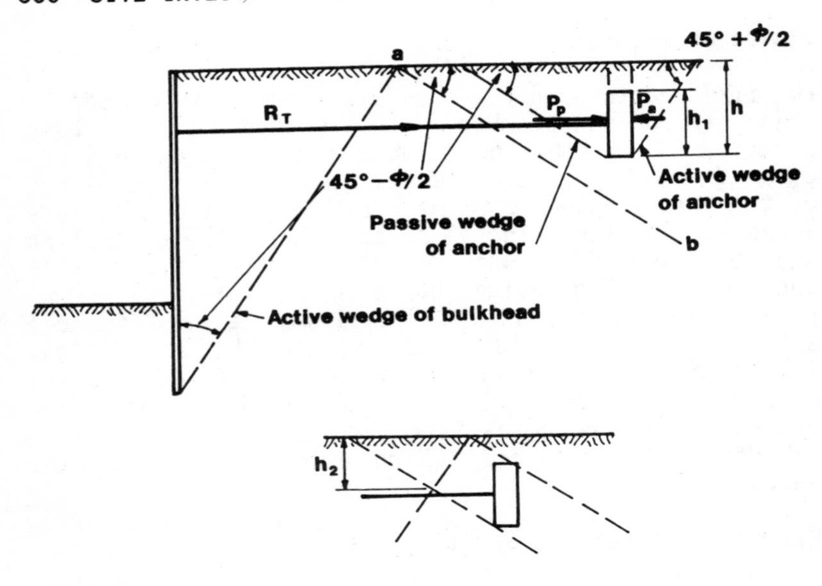

Figure 12.16 Rules for the design of deadman anchors.

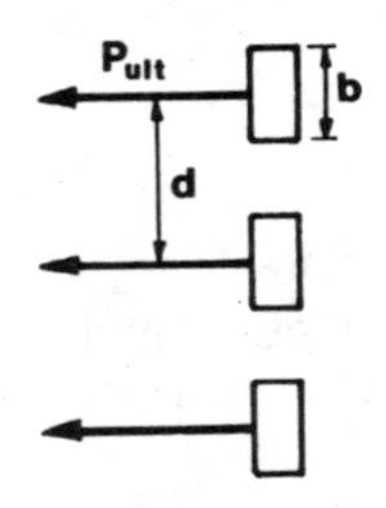

Figure 12.17 Separate anchors.

The resistance of a single anchor, (Figure 12.17) width b, height h_1 buried to depth h, where $h_1 > \frac{1}{2}h$ in a cohesionless soil is given by:

$$P_{ult} = {}^1/_2\gamma.h^2(K_p - K_a)b + {}^1/_3K_o(\sqrt{K_p} + K_a).h^3.\tan\phi$$

$$= R_T.b + {}^1/_3K_o(\sqrt{K_p} + K_a).h^3.\tan\phi$$

provided the calculated resistance per unit length of wall (P_{ult}/d) does not exceed R_T. In practice $P_{ult}/d < R_T$ provided $d > (b+0.1h)$, approx.

K_o is the coefficient of earth pressure at rest and may be taken as 0.4. For $h_1 < \frac{1}{2}h$, the anchors are treated as pad footings of width h_1, length b, and surcharge load $(h - \frac{1}{2}h_1)$, using Terzaghi's bearing capacity formula.

The same basic approach is used for cohesive soils but active and passive forces are calculated using Bell's solution or the Coulomb wedge method. For individual anchors the ultimate anchor resistance is

$$P_{ult} = b(P_p - P_a) + 2ch^2 \quad (\text{for } d > (b + 1/2h))$$

If the water table is above the anchor, the submerged density is used throughout. A factor of safety of at least 2 is used for anchor design.

The actual distribution of stresses on a retaining wall may be different from what is assumed, leading to an increase in tension in the tie rods. In addition, loading of the rod may occur through settlement of the surrounding soil, tending to cause sag in the rod and increase tension; and unequal yield of the wall may lead to some ties taking a greater proportion of the pull than others.

Because of these variations, rod design tension should be increased to 30% above calculated values for the rods themselves and to 50% above calculated for connections and splices.

THE STABILITY OF SLOPES

Stability problems can occur in natural slopes, in cuttings or in embankments. Failure can occur rapidly, with a large mass of material suddenly sliding down the slope; or more slowly, with gradual movement over a number of months, years or even decades. The most important single factor in slope stability is the pore water pressure. An increase in pore pressure reduces the effective stress - the pressure between soil grains - allowing them to slide over each other with less resistance. As a consequence, landslides tend to occur at wet periods of the year, or following prolonged heavy rain. Established slow moving landslides move at a faster rate during or just after wet periods.

Stability is also strongly influenced by zones of weakness in the material. This is particularly important in rock slopes, where a knowledge of the frequency and directions of such features as bedding planes, fissures and faults is vital to an understanding of the stability problem and to prediction of future stability. The stability of rock slopes is a specialist area, usually investigated by engineering geologists, and it is not covered here.

An analysis of slope stability is important for a number of reasons. Where it is proposed to construct an embankment or make a cutting, a suitable batter slope must be chosen. Too steep a slope will lead to instability, whilst too flat a slope will be excessively costly. The stability of an existing slope will be a crucial consideration before any decision can be made to develop an area. When designing remedial measures, the extent of material involved in a landslide or potential landslide must be assessed so that the extent of work required and its effectiveness can be judged.

Analysis of the stability of a slope consists of assuming a failure surface and comparing disturbing forces due to the weight of material and pore pressures with restraining forces provided by the shear strength of the slope material. The ratio of the maximum restraining force which can be developed along a potential slip surface to the amount actually required for stability gives the factor of safety against slope failure along that surface. Usually, a number of potential slip surfaces are analysed until the surface giving the minimum factor of safety is obtained.

The short-term stability of a cutting can be determined by using total stress soil strength parameters. However, the effect of pore pressures on the stability of slopes is of critical importance so the stability of embankments and natural slopes and the long-term stability of cuttings can be properly assessed only by the use of effective stress parameters, taking into account pore pressures developed within the slope. Generally, the long-term stability of a slope is the most critical and it is usual practice to carry out slope stability analyses in terms of effective stresses. Where possible, piezometers should be installed into existing slopes, and into major embankments as they are built, so that pore pressures used in the analyses accurately reflect actual site values.

MODES OF FAILURE

For purposes of analysis, slope failure may be divided into the categories listed below:

1. Plane slide
   (a) surface slide
   (b) subsurface slide
2. Deep seated slip
   (a) wedge failure
   (b) rotational failure
   (c) failure along an irregular surface

The different types of failure are illustrated in Figure 12.18.

Analysis of surface and shallow subsurface slides

Surface sliding occurs only in granular soils, for which c = 0. In cohesive soils the cohesion binds the particles together and makes deep-seated rotational failure more critical. In clay soils for which c'(the cohesion in terms of effective stress) is zero, subsurface sliding may manifest itself as a surface creep movement.

For the slice of material shown in Figure 12.19 the normal force on the base due to its weight is

$$N = W\cos\alpha$$

The pore pressure ratio r_u is defined as the ratio of pore water pressure to overburden pressure. Thus, at the base of the slide, pore water pressure is $r_u.W/ds$ and the normal thrust along the base (length $ds/\cos\alpha$) is

$$U = \frac{r_u.W}{ds} \quad x \quad \frac{ds}{\cos\alpha} = r_u.W\sec\alpha$$

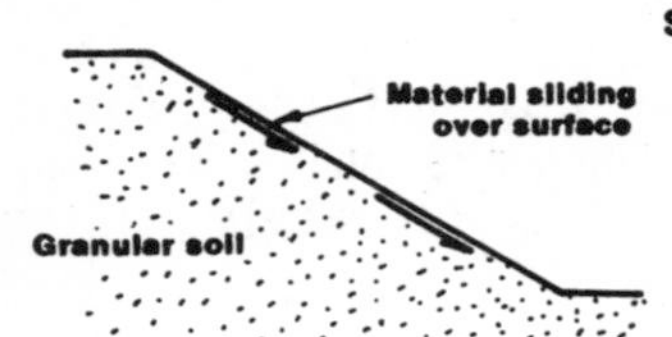

SURFACE SLIDING

Surface sliding occurs in granular materials. It is the most critical mode of failure where soil and groundwater conditions are homogeneous in a slope of granular material: i.e. the slope should be dry, fully saturated to the surface, or submerged.

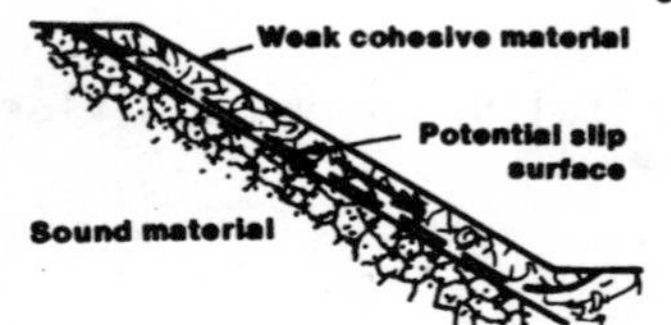

SHALLOW SUBSURFACE SLIDING

The presence of cohesion inhibits surface sliding so that deep-seated slips are more critical. However, a subsurface slide may occur at the base of a thin covering of weak material on a hillside.

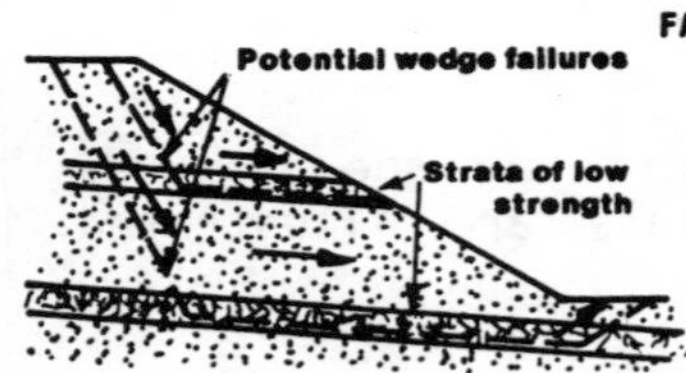

FAILURE ALONG AN IRREGULAR SURFACE

In a slope with complex soil and groundwater conditions, and perhaps an irregular face, failure or potential failure surfaces may not follow any of the standard patterns shown above.

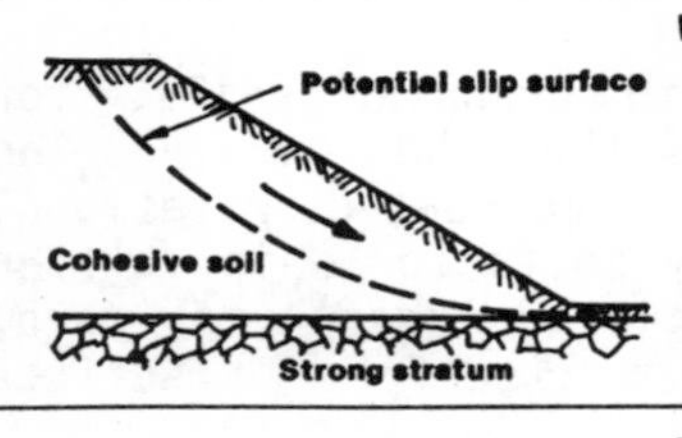

WEDGE FAILURES

Wedge failures usually occur where the hillside is made up of bands of strong and weak material. The positions of the failure wedges are controlled largely by the occurrence of the weak and strong bands.

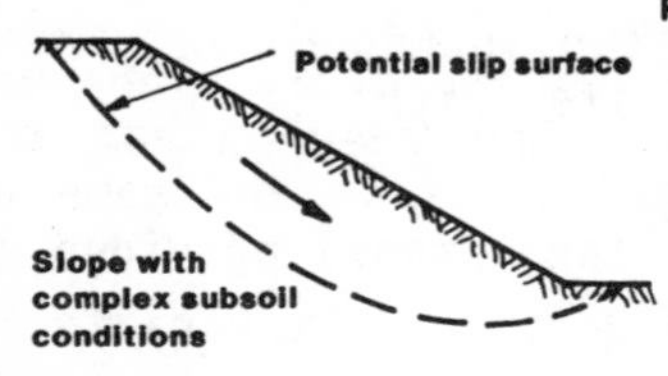

ROTATIONAL FAILURE

The most common form of failure in cohesive soils is deep rotational slip. The deepest surfaces which can form within the weak material are usually the most critical. Thus, the base of the critical slip surface is often limited by an underlying strong stratum.

Figure 12.18 Modes of slope failure.

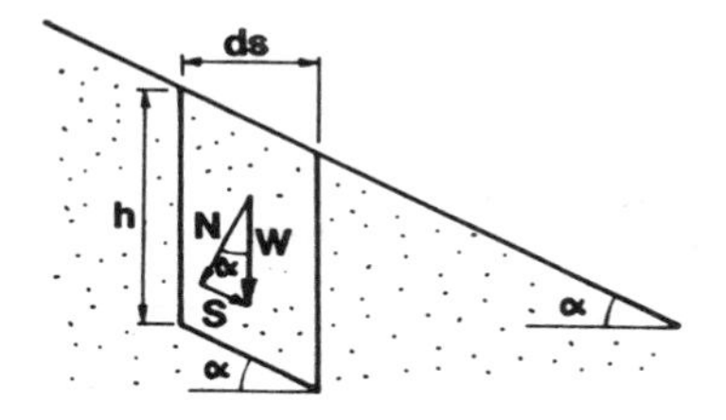

Figure 12.19 Forces on a slice of material.

If ϕ' is the effective angle of shearing resistance then the maximum resistance R due to sliding along the base of the slice is

$$R = (N - U)\tan\phi' = W(\cos\alpha - r_u\sec\alpha)\tan\phi'$$

The sliding force due to the weight of the slice is

$$S = W\sin\alpha$$

and the factor of safety F against sliding is defined as

$$F = \frac{R}{S} = \frac{W(\cos\alpha - r_u\sec\alpha)\tan\phi'}{W\sin\alpha}$$

$$= \frac{\tan\phi'}{\tan\alpha}(1 - r_u\sec^2\alpha)$$

For a dry slope $r_u = 0$. For a saturated slope, $r_u = \gamma_w/\gamma$ where γ is the bulk density of the material and γ_w is the density of water. For submerged slopes, the analysis can be carried out as for a dry slope not using submerged densities, with the result that the factor of safety for a submerged slope is the same as that for a dry slope (i.e. $F = \tan\phi'/\tan\alpha$) provided ϕ' remains unchanged.

A similar analysis can be carried out for a c-ϕ soil. In this case there is an additional restraining force C on the base of the soil due to the effective cohesion c', acting over the base of the slice. Thus

$$C = c'.ds.\sec\alpha$$

The factor of safety against sliding is now

$$F = \frac{R + C}{S}$$

$$= \frac{c'.ds.\sec\alpha}{W\sin\alpha} + \frac{\tan\phi'}{\tan\alpha}(1 - r_u\sec^2\alpha)$$

The weight of the slice can be expressed in terms of its volume and density. If the depth of the slice is h then

$$W = \gamma.h.ds$$

Hence

$$F = \frac{c'.ds.\sec\alpha}{\gamma.h.ds.\sin\alpha} + \frac{\tan\phi'}{\tan\phi}(1 - r_u.\sec^2\alpha)$$

$$= \frac{1}{\tan\alpha}\left[\frac{c'}{\gamma h}\sec^2\alpha + \tan\phi'(1 - r_u.\sec^2\alpha)\right]$$

From these calculations it can be seen that for a cohesionless material the factor of safety is independent of depth so that, in theory, sliding can take place at any depth. In practice, the material density, therefore its strength, is likely to increase with depth so that sliding will probably be a surface or shallow subsurface phenomenon. With cohesive soils, on the other hand, the factor of safety against sliding is infinite at the surface (where h = 0) and decreases with depth. This explains why slides in cohesive soils tend to form at the bottom of weak bands. It also demonstrates why deep-seated slips occur in cohesive soils, rather than shallow slides, unless there is a thin cover of weak material on a hillside, underlain by sound material.

Wedge analysis

Where distinct weak bands occur within a slope, sliding may occur along a series of plane or nearly plane slip surfaces as shown in Figure 12.20, so that the slipped material forms a series of wedges. Analysis is carried out by first assuming a failure surface and then considering the forces on the wedges due to their

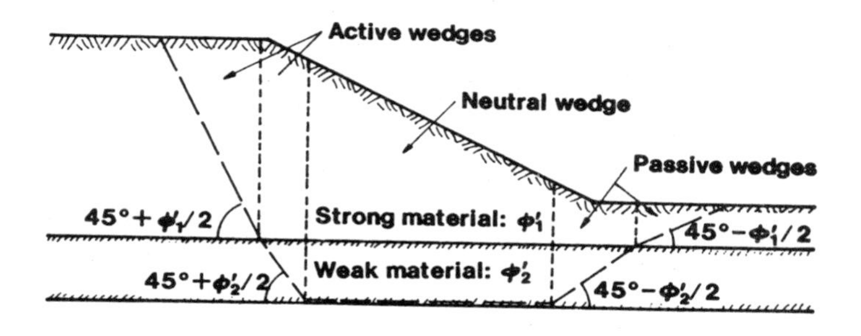

Figure 12.20 Example of an assumed wedge failure.

weights and the friction and cohesion forces both along their bases and between wedges. The trick is to find the slip surface which gives the lowest value of factor of safety because this is the one along which failure is most likely to occur. Figure 12.20 includes some rules which can be used when assuming the wedge shapes but, even using these, a number of trials may be required before the critical slip plane is located. This makes hand calculations lengthy and tedious: computers can be a great help in this work, especially if more than three wedges are involved. For hand calculations, the problem is usually solved graphically. For the two-wedge problem shown in Figure 12.21(a), the forces on the wedges and their directions are drawn in, as shown in Figure 12.21(b). These are:

- weights W_1 and W_2, which can be obtained from the area of each wedge and the bulk density of the soil,
- cohesion forces C_1 and C_2, where

$$C_1 = \frac{c_1}{F} \times \text{length AF},$$

$$C_2 = \frac{c_2}{F} \times \text{length FE, and}$$

F is the factor of safety against wedge failure,
- thrusts U_1 and U_2 due to pore water pressures, which are calculated in the usual manner,

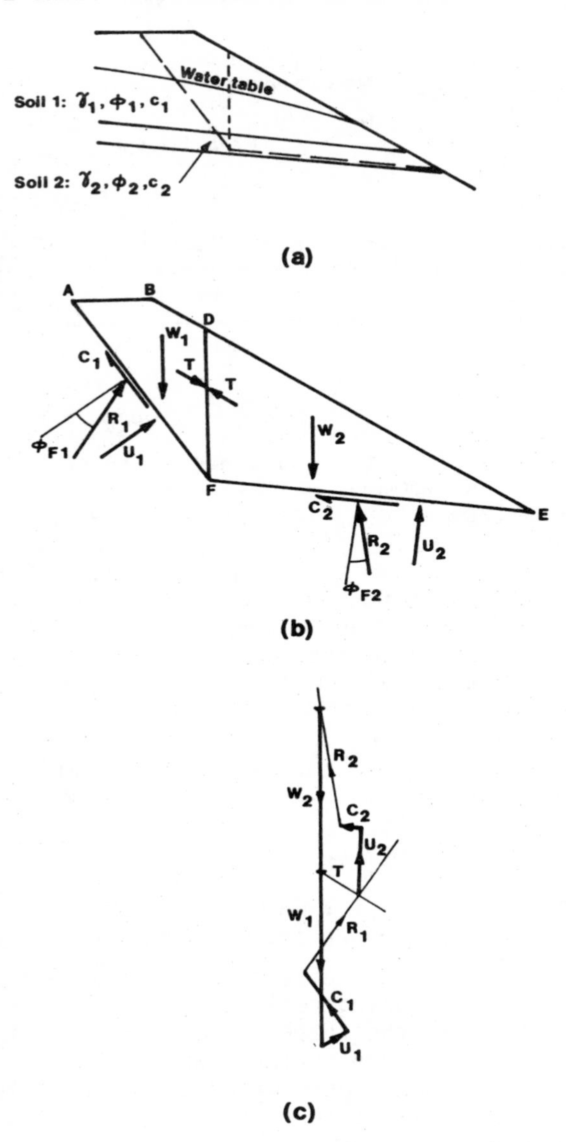

Figure 12.21 Analysis of a two-wedge problem: (a) slope conditions and assumed failure surfaces; (b) forces acting on the wedges; (c) construction of the force diagram.

- reactions R_1 and R_2, whose values are unknown, but whose directions ϕ_{F1} and ϕ_{F2} can be calculated from $\tan\phi_{F1} = \tan\phi'_1/F$, $\tan\phi_{F2} = \tan\phi'_2/F$, and
- interslice thrust T, about which nothing is known. This causes the problem to be statically indeterminate and an assumption about the direction of T must be made. T is usually assumed to act either at an angle ϕ to the horizontal, or parallel to the slope. In practice, the choice of direction has little influence on the calculated factor of safety.

A factor of safety F is first guessed and then a polygon of forces is drawn as shown in Figure 12.21(c). When the polygon closes, the correct factor of safety has been chosen.

Rotational failure

This is probably the most common form of failure. Analysis is usually carried out assuming the failure surface follows a circular arc as indicated in Figure 12.22. The usual approach used is that proposed by Bishop. The factor of safety is defined as

$$F = \frac{\text{Available shear resistance of the soil, } \tau}{\text{Shear resistance actually mobilised, } s}$$

The available shearing resistance at any point on the slip surface is given by the Mohr-Coulomb failure criterion:

$$\tau = c' + (\sigma - u)\tan\phi'$$

where c' and ϕ' are the effective cohesion and angle of shearing resistance,
σ is the total normal stress on the failure surface, and
u is the pore water pressure at the failure surface

The shear strength s, actually mobilised, will be some proportion of this:

$$s = \frac{c' + (\sigma - u)\tan\phi'}{F}$$

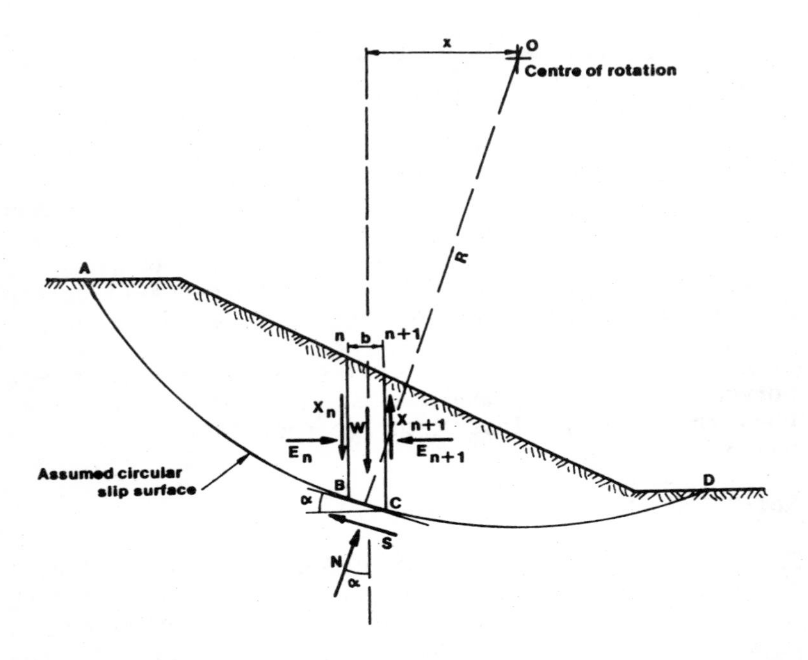

Figure 12.22 The mechanics of circular slip using the Bishop method of slices.

the assumption here being that the factor of safety F can be applied equally to the cohesion and shearing resistance.

The procedure to determine the factor of safety is to take moments about the centre of rotation, O, of the slip circle. The factor of safety then becomes

$$F = \frac{\text{(maximum force developed along arc ABCD).R}}{\text{(weight of material).(horizontal distance of centre of gravity from O)}}$$

In practice the shear stress varies throughout the length of arc ABCD and depends on the normal stress and pore water pressure at any point along the slip surface. To overcome this problem the slope is divided into slices and the forces on each slice are considered individually as indicated in Figure 12.22. The resulting expression for the factor of safety is

$$F = \frac{R}{\sum W.x} .\sum[c'L + (N - uL)\tan\phi']$$

where R is the radius of the assumed slip circle
W is the weight of each slice
x is its distance from the centre of rotation ($x=\sin\alpha/R$)
L is the length of the base of the slice ($L = b.\sec\alpha$, where b is the slice width)
N is the normal force on the base of the slice.

We are now faced with the problem, that, unfortunately, thrust N depends on the interslice forces E and X, which are unknown. A solution is obtained by simply sweeping these terms under the carpet and ignoring them. Thus, resolving vertically $N = W.\cos\alpha$, to give

$$F = \frac{1}{\sum W.\sin\alpha} \cdot \sum[c'L + \tan\phi'(W.\cos\alpha - uL)]$$

This equation allows the factor of safety to be calculated directly and is suitable for hand calculations or simple computer programs. Values of F obtained in this way are conservative (that is, on the low side of the true factor of safety) especially if the base angle, α, is high in a number of slices.

Hand calculations for a single slip circle are tedious and for most problems a number of circles must be analysed to find the one that gives the minimum factor of safety. Slip circle analysis is therefore almost always carried out using a computer program. Once the decision is made to use a computer for analysis, the need to simplify the basic equation is less pressing. A more rigorous solution can be obtained by considering the effect of the interslice forces on the base thrust N. To solve the resulting equation rigorously requires assumptions to be made about the distribution of interslice forces. The process is complex, even for a computer application, and the value of F obtained depends on the assumptions made. However, the problem can be overcome by ignoring the shear forces ($X_n - X_{n+1}$) across each slice, to give

$$F = \frac{1}{\sum W.\sin\alpha} \cdot \sum\left[\frac{\{c'b + \tan\phi'(W-ub)\}\sec\alpha}{1 + \frac{1}{F}\tan\phi'.\tan\alpha}\right]$$

This equation gives considerably more accurate values of F. Because F appears on both sides of the equation, it is solved by making an initial guess at its value and solving by successive iterations, using the value of F calculated from one calculation as the guess for the next calculation. An initial guess of F=1 or F=2 is usually used. The solution usually converges to within 0.01F in about 3 iterations.

Where a slope is partly submerged, as illustrated in Figure 12.23, it is partly supported by the water

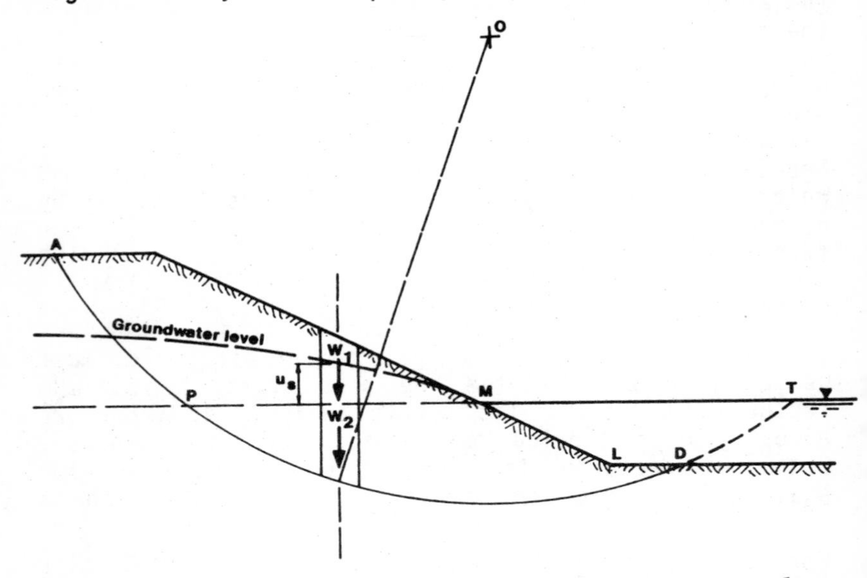

Figure 12.23 Circular slip in a partially submerged slope.

pressure acting on it. The total disturbing moment of the material contained within the slip is reduced by the moment of the water pressure acting on the surface DLM, in the figure. This will be equal to the weight of water bounded by MLDT. Note that the boundary of the body of water being taken into account must follow the slip surface so that the water pressure acting on the boundary has no moment about O and does not therefore need to be considered. If, for instance, the weight of water bounded by MLDU were considered, the water pressure on UD would have a moment about O, so it would have to be taken into account. In effect, the water can simply be considered as an extra material with weight but no shear strength and calculations can be carried out as before but with the slip circle extended to T. An alternative treatment, suggested by Bishop to aid computation, is to consider the whole body of water bounded by segment PMTDP. Since PT is a horizontal line, this body of water will be symmetrical about a vertical line through O and will have no moment about O, so its effect can be ignored in calculations. The weight of water in area PMLDP can be deducted from the calculations by taking the submerged density of the soil below the free water surface (line PT). However, using submerged weights below line PT means that pore pressure levels up to this line have already been allowed for in calculations so that any pore pressure heads in slices whose bases are below the external water level are measured from line PT, as indicated by u_s in Figure 12.23. Thus, in the equations for the factor of safety:

- W is replaced by $(W_1 + W_2)$ where W_1 is the weight of the slice above the external water table and W_2 is the submerged weight below it,
- u is replaced by u_s, the pore pressure measured from the level of the external water surface, for submerged or partially submerged slices, and
- the external water pressure can then be ignored.

Failure along an irregular surface

For the majority of slope stability problems, slip surfaces are circular or nearly circular and the slip circle analysis described above is adequate. Occasionally, where there are weak bands of material below a slope, the failure or potential failure surface may be substantially non-circular, as illustrated in Figure 12.24. This problem has been analysed by Janbu, using a method of slices similar to that used in the Bishop slip circle method but without the assumption of a circular slip surface. The approach is similar to that used for a wedge analysis but with a large number of thin wedges. The resulting approximate solution,

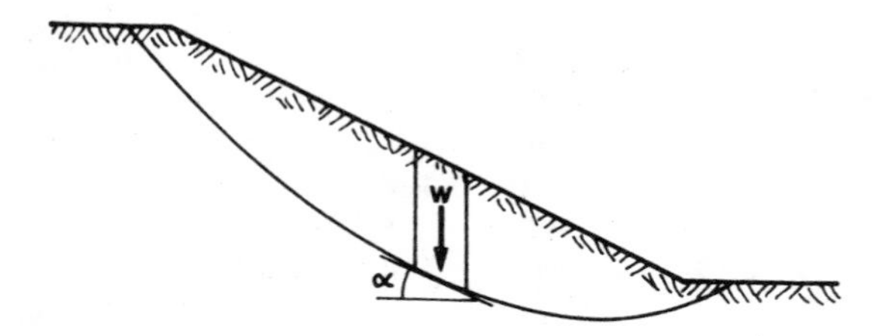

Figure 12.24 Analysis of slip along an irregular surface using the Janbu method.

ignoring interslice shear forces is

$$F = \frac{1}{\sum W.\sin\alpha} \cdot \sum\left[\frac{\{c'.b + \tan\phi'(W-ub)\}\ \sec\alpha}{1 + \frac{1}{F}\tan\phi'\ \tan\alpha}\right]$$

which is identical to the second approximation for the Bishop solution. As with the Bishop method, more accurate solutions exist, which rely on assumptions about interslice forces, but they are not usually used.

Specifying the slip surface is less simple for non-circular slips. Usually it is specified as a series of co-ordinates. This means quite a lot of data input is required if a number of surfaces are to be analysed. Also, deciding on what shapes to choose can itself be a problem. By contrast, circular slip surfaces are usually specified for computer input simply by giving the co-ordinates of the centre of rotation and the level of the lowest point of the circle.

Standard solutions - slope stability curves

The stability of simple slopes with homogeneous soil conditions is usually obtained using slope stability curves produced by Bishop and Morgenstern. These were produced from the analysis of a large number of slopes using the Bishop method. The factor of safety of the slope shown in Figure 12.25 can be expressed in terms of stability coefficients m and n and the pore pressure ratio r_u:

$$F = m - n.r_u$$

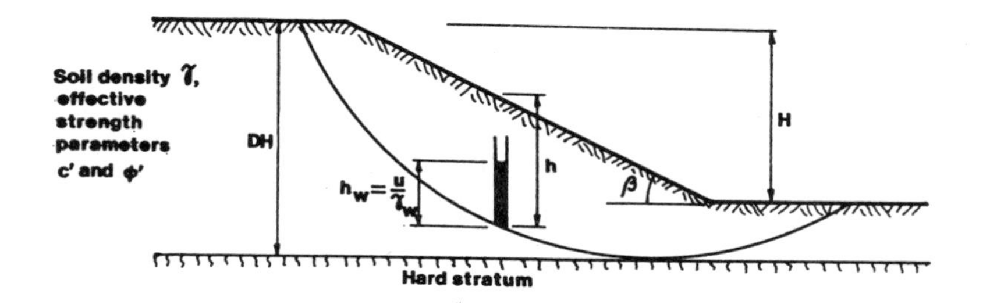

Figure 12.25 Definition of the parameters used in the Bishop and Morgenstern slope stability charts.

where

m and n are obtained from Figure 12.26;
r_u is the pore pressure ratio; the ratio of pore water pressure, u, to overburden pressure, γh. Thus

$$r_u = \frac{u}{\gamma h} = \frac{\gamma_w h_w}{\gamma h}$$

The first stage is to calculate the value of the dimensionless number c'/H, where H is the height of the slope. Sets of charts of m and n are given for values of

$$\frac{c'}{\gamma H} = 0.025 \text{ and } 0.050 \text{ .}$$

In order to select the appropriate chart, a suitable value of depth factor D must be chosen (see Figure 12.25). For a slope founded on a strong base, D = 1 but where the base is composed of similar material to the slope, it is not obvious which value of D will give the minimum factor of safety. Bishop and Morgenstern provide a method of determining the appropriate depth but the procedure is rather cumbersome and in practice it is easier simply to determine values of F for the whole range of depth factors and take the lowest value.

In using these curves, the following points should be noted.

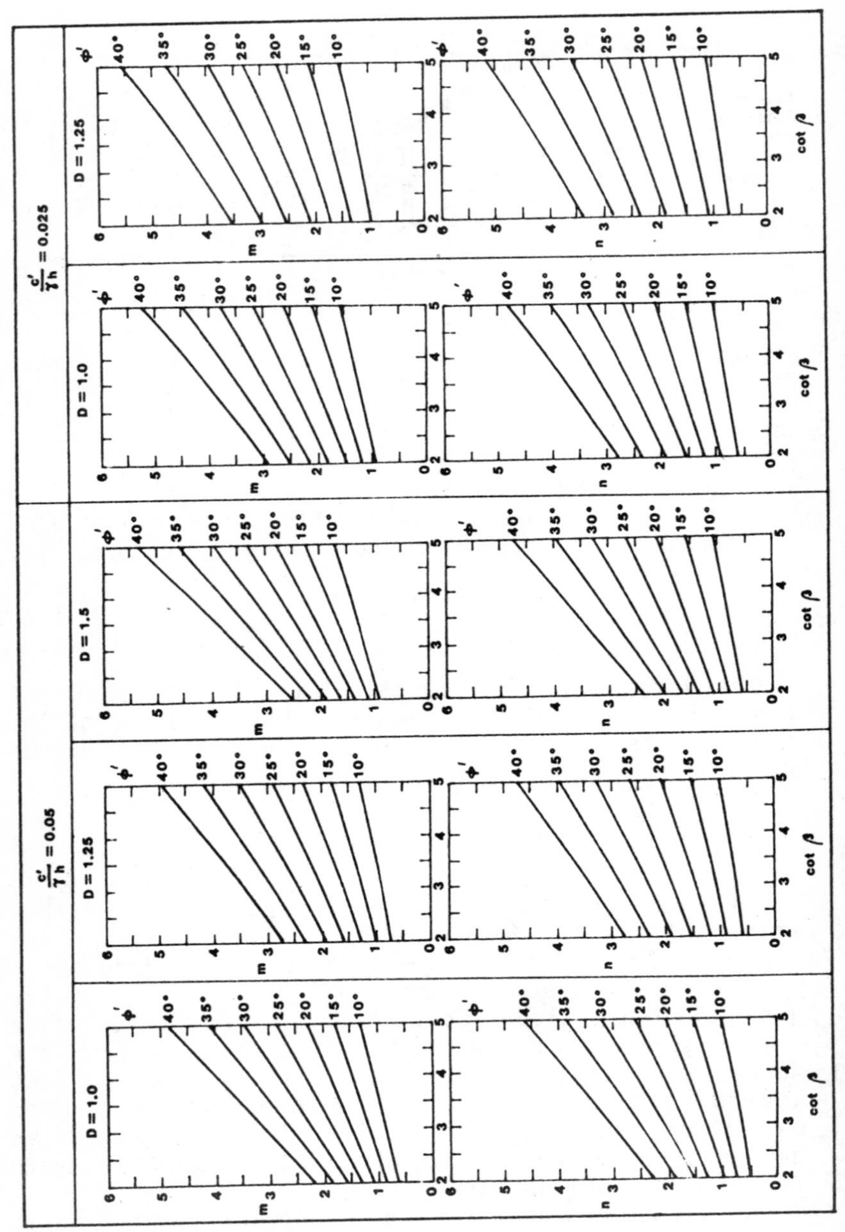

Figure 12.26 Slope stability charts by Bishop and Morgenstern.

(1) In determining which value of D to use, the presence of any hard stratum must be taken into account.

(2) For $c'/\gamma H = 0.025$, the value of $D = 1.5$ seldom gives a more critical factor of safety, so these charts have been omitted.

(3) For $c'/\gamma H = 0$ the surface slide condition is critical and the factor of safety should be determined from the expressions for surface slides, given earlier.

REFERENCES AND BIBLIOGRAPHY

BELL, F.G. (Editor) "Site investigations in areas of mining subsidence", Newnes-Butterworth, (1975).

BELL, F.G. (Editor) "Foundation engineering in difficult ground", Newnes-Butterworth, (1978).

BEREZANTSEV, V.G. "Load bearing capacity and deformation of piled foundations", Proceedings of the 5th International Conference on Soil Mechanics and Foundation Engineering, Paris (1961).

BISHOP, A.W. " The use of the slip circle on the stability analaysis of slopes", Géotechnique, Vol. 5 (1955).

BISHOP, A.W. and N. MORGENSTERN "Stability coefficients for earth slopes", Géotechnique, Vol. 10 (1960).

BOWLES, J.E. "Foundation analysis and design", McGraw-Hill, 3rd ed. (1982).

BRITISH STANDARD 1377 "Methods of test for soil for civil engineering purposes", British Standards Institution (1975).

BRITISH STANDARD 5930 "Code of practice for site investigations", British Standards Institution (1981).

BROMHEAD, E.N. "The stability of slopes", Blackie and Son Ltd. (1986).

BROMS, B. "Methods of calculating the ultimate bearing capacity of piles, a summary", Sols-Soils, Vol. 5 (1966).

BRUHN, R.W., M.O. MAGNUSON and R.E. GRAY "Subsidence over abandoned mines in the Pittsburgh Coalbed", Proceedings of 2nd International Conference on Ground Movements and Structures, Cardiff, Pentech Press (1980) 142-156.

BUILDING RESEARCH ESTABLISHMENT "Concrete in sulphate-bearing soils and ground waters", Digest 250, BRE (1981).

BURLAND, J.B. and C.P. WROTH "Review paper: settlement of buildings and associated damage", Proceedings of the Conference on Settlement of Structures, Cambridge, Pentech Press (1974) 611-654.

CAQUOT, A and J. KERISEL "Tables for the calculation of passive pressure, active pressure and bearing capacity of foundations", Gauthier-Villars.

CARTER, M. "Geotechnical engineering handbook", Pentech Press (1983).

CARTER, M. and S.P. BENTLEY "Practical guidelines for microwave drying of soils", Canadian Geotechnical Journal, 23 (1986) 598-601.

CHIN, F.K. "Estimation of the ultimate load of piles not carried to failure", Proceedings of the 2nd South East Asian Conference on Soil Engineering, Singapore (1970) 81-90.

CHIN, F.K. "The inverse slope as a prediction of the ultimate bearing capacity of a pile", Proceedings of the 3rd South East Asian Conference on Soil Engineering, Hong Kong (1972) 83-91.

CHIN, F.K. "Diagnosis of pile condition", Journal of Geotechnical Engineering, Asian Institute of Technology, 9 (1978) 85-104.

DUMBLETON, M.J. and G. WEST "Preliminary sources of information for site investigations in Britain", Department of the Environment RRL Report LR403 (1976).

GEOTECHNICAL CONTROL OFFICE, HONG KONG "Geotechnical manual for slopes", Government Publications Centre, Hong Kong, 2nd ed (1974).

GIROUD, J.P. "Stresses under linearly loaded rectangular area", Journal of the American Society of Civil Engineers, Vol. 96, SM1 (1970).

HANSON, J.B. "A general formula for bearing capacity", Danish Geotechnical Institute Bulletin No. 11 (1961).

HANSON, J.B. "A revised extended formula for bearing capacity", Danish Geotechnical Institute Bulletin No. 28 (1968).

HANSON, J.B. "Code of practice for foundation engineering", Danish Geotechnical Institute Bulletin No. 32 (1978).

HEAD, K.H. "Manual of soil laboratory testing", Pentech Press:

Vol. 1 "Soil classification and compaction tests" (1980).

Vol. 2 "Permeability, shear strength and compressibility tests" (1982).

Vol. 3 "Effective stress tests" (1985).

HEALY, P.R. and J.M. HEAD "Construction over abandoned mine workings", CIRIA/PSA Special Publication 32 (1984).

HIGGINBOTTOM, I.E. "Methods of development above ancient shallow pillar-and-stall coal workings", Proceedings of 2nd International Conference on Reclamation, Treatment and Utilisation of Coal Mining Wastes, Nottingham, Elsevier (1987) 639-652.

KIRKDAM, R. "Discussion", Proceedings, Brussels Conference 58 on Earth Pressure Problems (1958).

LAMB, T.W. and R.V. WHITMAN "Soil Mechanics, John Wiley and Sons (1979).

MEYERHOF, G.G. "The ultimate bearing capacity of foundations", Géotechnique, Vol. 2, No. 4 (1951).

NATIONAL COAL BOARD "Subsidence engineers' handbook", Coal Board Mining Department, London, 2nd ed. (1975).

NEWMARK, N.M. "Influence charts for computation of stresses in elastic soils", University of Illinois Engineering Experimental Station Bulletin No. 338 (1942).

NORDLAND, R.L. "Bearing capacity of piles in cohesionless soils", Proceedings of the American Society of Civil Engineers, 89, SM3 (1963) 1-35.

PECK, R.B., W.E. HANSON and T. THORNBURN "Foundation engineering", John Wiley and Sons, 2nd ed. (1974).

PIGGOTT, R.J. and P. EYNON "Ground movements arising from the presence of shallow abandoned mine workings", Proceedings of Conference on Large Ground Movements and Structures, Cardiff, Pentech Press (1977) 749-780.

POULOS, H.G. and E.H. DAVIES "Elastic solutions for soil and rock mechanics", John Wiley and Sons (1974).

RICS "The problems of disused mine shafts", The Royal Institution of Chartered Surveyors, Minerals Division (1978).

ROWE, P.W. "Cantilever sheet piling in cohesionless soil", Engineering, 172 (1951).

SHADBOLT, C.H. "Mining subsidence; historical review and state of the art", Proceedings of Conference on Large Ground Movements and Structures, Cardiff, Pentech Press (1977) 705-748.

SKEMPTON, A.W. and L. BJERRUM "A contribution to the settlement analysis of foundations on clay", Géotechnique, Vol. 7 (1957).

SMITH, G.N. "Elements of soil mechanics for civil and mining engineers", Granada Publishing, 5th ed. (1982).

SYMONS, M.V. "Sources of information for preliminary site investigations in old coal mining areas", Proceedings of Conference on Large Ground Movements and Structures, Cardiff, Pentech Press (1977) 119-135.

SYMONS, M.V. "Site investigation in old coal mining areas; recommended procedure for the desk study", Proceedings of the Third International Conference on Ground Movements and Structures, Cardiff, Pentech Press (1984) 173-187.

TAYLOR, R.K. "Site investigations in coalfields; the problem of shallow mine workings", Quarterly Journal of Engineering Geology I (1969) 115-133.

TERZAGHI, K. and R.B. PECK "Soil mechanics in engineering practice", John Wiley and Sons, 2nd ed. (1967).

TOMLINSON, M.J. "Pile design and construction practice", Viewpoint Publications (Cement and Concrete Association) (1977).

TOMLINSON, M.J. "Foundation design and construction", Pitman, 4th ed. (1980).

TOMLINSON, M.J., R. DRISCOLL and J.B. BURLAND "Foundations for low-rise buildings", The Structural Engineer, Part A, 56A (6) (1978) 161-173: reprinted as CP61, Building Research Establishment (1978).

UNITED STATES STEEL "Steel sheet piling design manual" USS (1975).

U.S. NAVY "Design manual: soil mechanics, foundations and earth structures", Navfac DM-7, Department of the Navy, Naval Facilities Engineering Command (1971).

WELTMAN, A.J. and J.M. HEAD "Site investigation manual", CIRIA/PSA Special Publication 25 (1983).

WHITAKER, T. "Experiments with model pile groups", Géotechnique, Vol. 7 (1957).

WHITAKER, T. "The design of piled foundations", Pergamon Press, 2nd ed. (1976).

WINTERKORN, H.F. and H.Y. FANG "Foundation engineering handbook", Van Nostrand Reinhold (1975).

INDEX

Abandonment plans 169, 171
Acidity test 103
Active earth pressure 286-293
Adhesion
  and friction at interfaces 288, 289
  on piles 265-269
Adsorption complex 224
Aeolian soils 109
Aerial photographs 45, 46, 48, 175, 177, 211
Aerial photography 45, 46
Air voids content 95
Alcohol method for moisture content 81
Allowable bearing pressure 4, 5, 127, 137, 138, 233, 244, 261, 262
Alluvial soils 108
Anchorages 34, 305-307
Anchored bulkheads 33, 34, 300-307
Angle of draw
  longwall working 198
  pillar and stall working 193, 194
ASTM classification system 118
Atterberg limit tests 78, 81-83

Balloon density apparatus 97, 98
Bearing capacity
  factors 236-241
  see Allowable bearing pressure, Ultimate bearing capacity, Bearing capacity factors
Bearing failure 209
Bearing pressures 126
  allowable 4, 5, 127, 137, 138, 233, 244, 261, 262
  rigid retaining walls 296
  ultimate 4, 126, 127, 137, 138, 227, 233-241, 244, 245
Bell pits 163
Bell's solution 292, 296
Bill of quantities 49
Bishop analysis for slopes 312-319
Black cotton soils 160
Borehole logs 121, 124, 125, 183, 184
Boreholes and trial pits
  common methods 53-60
  for foundations 52
  for roads 52
  for pipelines 52
  for runways 52
  in mining areas 179-183
  in rock 58, 59
  in soil 54-57
  numbers and depths required 49, 52, 53, 182-184
  post hole auger 54
  rotary auger 54, 55
  rotary coring 58, 59
  shell and auger 56, 57
Braced excavations 297, 298
BS classification system 118, 119, 121-125
Bulb of pressure 15, 246, 247

California bearing ratio 79, 99, 100
Cantilever retaining walls 33, 34, 298-300
Capping beam 23
Casagrande classification system 118
Casagrande liquid limit device 82
Cement content for foundations 128
Chemical
  attack 128, 129, 145 160
  tests 79, 103, 104
Chin plot 280-283
CLASP system 204
Classification of soils 109, 117-123
Coefficient of
  consolidation 79, 92, 124, 135, 228-232, 259, 260
  earth pressure 285-293
  earth pressure on piles 267
  volume compressibility 79, 92, 115, 116, 124, 135, 150, 228-232, 252, 254-257, 259, 261
Collapsible soils 157
Colluvial soils 109
Compaction 78, 93, 100
Compactness of sands 110-112
Compressibility, coefficient of volume 79, 92, 115, 116, 124, 135, 142, 150, 228-232
Compressible soils 149-153
Consolidation
  and settlement 2, 5, 13, 92, 137, 138, 148-153, 159, 210-213, 225, 228, 252, 254
  coefficient of 79, 92, 124, 135, 228, 231, 232, 259, 260
  test 78, 90-93, 124, 157
  see also Settlement

Constant head permeability test 100-103
Core cutter density test 98, 99
Cracking in buildings 207-213
Creep movements 4
Crown holes 166

Deep compaction 151
Deep stabilisation 151
Degree of saturation 217
Degree of weathering 116, 117
Density
bulk 93, 218
dry 93, 218
maximum dry 95
natural 78
relative 110-112
saturated 218
submerged 219
tests 96-99
Descriptions
of soils 108-117
of rocks 120
Desert soils 160
Design
recommendations 136-141
values for soil parameters 133-136
Desk study 37-47
for mining areas 169-179
Diaphragm walls 34, 35
Differential settlement 258, 259, 263, 269
see also Consolidation, Settlement
Dipmeter 70-72
Drainage 41, 155
Dynamic cones 63, 65
Dynamic pile testing 283

Earth pressure at rest 267, 285
Earth retaining walls
see Retaining walls
Eccentric loads on foundations 241-243
Echo soundings 75, 76
Effective stresses 223-226
Elastic solutions 245-258
Electricity supply 41
Embankments 139
see also Fill material
Engineering News formula 273-275
Eolian soils 109
Evaporites 108, 109
Excavations 139
Expansive soils 3, 5, 21, 143, 144, 153-157, 160, 213, 214

Factor of safety 4, 233, 245, 263, 269, 294
Falling head permeability test 100-103
Field density 63-70, 93, 103
Fill material 147, 159, 160
Fireclay 168
Fixed earth support method 302-305
Flooding 48
Foundations
basements and deep spread 24, 151, 156
caissons 24-27
cement content for 128
eccentric loads on 241-243
failures of 209
in mining areas 187-192, 196, 203-205
pad 23, 136, 137
piled (see Piles)
raft 20, 21, 138, 151, 187, 190, 195, 203
size of 20
stability of 227
strip 17, 19, 136, 137, 154
thickness of 20
types 16-33
Founding depth 12, 127, 136
Free earth support method 301, 302, 305
Frost susceptibility 145, 146, 159

Gas hazards due to mining 185
Gas supply 41
Geological maps 42, 43, 47, 212
in mining areas 169-177, 181
Geological memoirs 43, 169, 175
Geophysical surveys
echo soundings 75, 76
gravimetric 73
in mining areas 179, 180, 184
magnetic 74
resistivity 72, 73
seismic 74, 75
Geotechnical reports 105-108
Glacial soils 109
Goaf or gob 197
Grading 78, 84, 85
Gravimetric surveys 73
Ground anchors 24
Groundwater
levels 6
pressures 6
Group efficiency ratio 270, 271
Grouting old mineworkings 185, 186, 192-195

Handbooks of British Regional Geology 43, 44
Harmonic extraction 206
Hiley formula 275-278

Identification and classification of soils 108-123
In-situ testing 63-70, 93, 103
Isometric plotting 132, 133

Janbu analysis for slopes 319, 320

Laboratory test schedules 77
Laboratory tests
California Bearing Ratio 79, 99, 100
chemical 79, 103, 104
compaction 78, 93, 100
consolidation 78, 90-93, 124, 157
density 78, 93, 96-99, 110-112
grading 78, 84, 85
liquid limit 78, 81-83
linear shrinkage 144
moisture content 78, 80, 81
permeability 78, 93, 100-103
plastic limit 78, 81-83
plasticity index 78, 81
reporting of 77
shear strength 67, 68, 78, 79, 86-90, 125, 126, 139 (see also Triaxial, Shear box, etc.)
specific gravity of solids 78, 83
Landslides 48
see also Slope stability
Lateral earth pressures 267, 284-293

Liquefaction 158
Lime columns 151
Liquid limit 78, 81-83, 143
Liquidity index 144, 145
Loess 109
Longwall mining 164, 165, 196-198

Mackintosh probe 66
Magnetic surveys 74
Maps
  geological 43, 47, 212
  in mining areas 169-177, 181
  land use 44
  mineral resources 45
  pedological 43, 44
  topographical 42, 211, 212
Maximum dry density 95
Methylated spirit method for moisture content 81
Microwave oven drying for moisture content 81
Mine
  adits and entries 163, 165, 179-181, 185, 186
  entry register 174, 175
  shafts 165, 185
Mining
  areas (see Mining areas)
  bell pits 163
  crown holes 166
  gas hazards 185
  partial extraction 163, 164
  pillar and stall methods 163, 164
  pillar crushing 168
  pillar punching 167, 168
  pillar robbing 164, 183
  roof failure 166, 195
  solution 208
  subsidence problems 165-169, 198-203, 205-206, 208
  total extraction (longwall) methods 164, 165, 196-198
  void migration 166, 179, 183
Mining areas
  construction precautions in 184-196, 203-206
  safe and unsafe zones in 187-191
  site investigations in 169-183
Model soil sample 217
Mohr circle 126, 220
Mohr-Coulomb failure criterion 223, 288, 290
Moisture barriers 155, 156
Moisture content
  alcohol (methylated spirit) method 81
  definition of 80, 217
  microwave oven method 81
  natural 78, 80, 144
  oven drying method 80
  sand bath method 80
  seasonal changes 5, 6, 21
  Speedy tester 81
  value in testing schedules 78

Newmark charts 249, 251-254
Nuclear density meter 99

Open-drive sampler 56, 59-61
Optimum moisture content 95
Organic soils 109, 145
Organic matter content test 103

Particle size distribution 78, 84, 85
Passive earth pressure 286-293
Pavement design 140
Peat 109
Permeability tests
  field 69, 70, 93, 103
  laboratory 78, 100-103
pH test 103
Photogrammetry 45
Photographs
  during site visit 48
  of trial pits 53
Piezometers 70-72
Pile driving formulas 273-278
Piles 27, 129
  adhesion and skin friction 28, 29, 265-267
  allowable working load 269
  bored 28-31, 33
  carrying capacity 263-269
  Chin plot 280-283
  contiguous 35
  driven 28-32
  end bearing 28, 29
  group effect 263, 270
  in mining areas 196, 204
  load tests 278-283
  negative skin friction 263, 272, 273
  secant 35
  short bored 21, 23, 155
Pillar
  and stall workings 163, 164
  punching and crushing 168
  robbing 164, 183
Piston sampler 62
Plastic limit 78, 81-83
Plasticity
  chart 143
  index 78, 81
  of soils 114, 142, 143
  tests 78, 81-83
Plate bearing test 69
Poisson's ratio 252
Pore water pressures 139, 223-226
Porosity 217
Post hole auger 54
Pressuremeter 66, 67
Pre-wetting of foundations 156
Principal stresses 219, 220, 246
Probes 63, 65, 66

Reconnaissance 36, 37, 39, 47, 48, 50, 51
Recovery 120
Red coffee soils 160
Regolith 108
Relative density 110-112
Reports, geotechnical 105-108
Residual soils 108-160
Resistivity surveys 72, 73
Retaining walls
  anchorages 33, 34, 305-307
  anchored bulkheads 33, 34, 300-307
  braced excavations 297, 298
  cantilever 33, 34, 298-300
  diaphragm 33, 34
  earth pressures on 284-293
  rigid gravity 33, 34, 294-297
  stability calculations 293-295, 298-305
  types 33, 34, 284

Road pavement design 140
Rock
  anchors 24
  descriptions 120
  quality designation 120
Rotary auger boring 54, 55
Rotary coring 58, 59

Samplers
  open-drive 56, 59-61
  piston 62
  split spoon 63, 64
Samples
  bulk 59, 60
  care and labelling 60, 61
  undisturbed 56, 59-62
  water 60
Sampling 59-63
  frequency 59
Sand bath method for moisture content 81
Sand drains 151
Sand replacement density test 96, 97
Saturation, degree of 217
Seasonal moisture changes 5, 6, 21
Seatearth 167
Sedimentation analysis 84, 85
Seismic surveys 74, 75
Sensitive clays 157, 158
Sensitivity 158
Set, of piles 274-276
Settlement calculations
  corrections 258, 259
  quick estimates 260, 261
  rate of settlement 259, 260
  rigid retaining walls 296
  using consolidation theory 254-258
  using elastic theory 252-254
Settlement 2, 5, 13, 92, 137, 138, 148-153, 159, 210-213, 225, 228, 233, 258, 259, 263, 296
  see also Consolidation, Settlement calculations
Sewerage 41
Shallow slides 309-312
Shear box test 79, 89, 235
Shear strength
  basic considerations 219-223, 228
  descriptions 110-113, 142
  design values 133, 134
  tests (see Laboratory tests, Triaxial, Shear box, etc.)
Shear stresses in soil 226
Shear vane test 79
Shell and auger boring 56, 57
Shrinkage/swell problems 3, 5, 213, 214
Sieve analysis 84, 85
Site investigation reports 11, 12, 210
Site investigations
  approaches and access 40
  depth of 8, 14-16
  desk study 38-47
  information required from 36
  land survey 40
  phases 36, 37, 49
  planning 8, 37-39
  reconaissance 36, 37, 39, 47, 48, 50, 51
  restrictions on land use 40
Site visits 36, 37, 39, 47, 48, 50, 51
Skin friction on piles 28, 29, 265-267
Slip circle analysis 315-319
Slope stability 52, 139, 308, 309
  circular slips 315-319
  non-circular slips 319, 320
  shallow slides 309-312
  stability curves 320-323
  wedge analysis 312-315
Slopes, natural 139
Soil
  classification 109, 117-123
  descriptions 108-117
  structure 111, 113
Soil-structure interaction 4
Solution mining 208
Specific gravity of soil solids 78, 83
Specifications 49
Speedy moisture content tester 81
Split spoon sampler 63, 64
Stability
  curves for slopes 320-323
  of foundations 227, 235
  of retaining walls 293-295
  of slopes, embankments and excavations 227, 228, 309-323
Standard penetration test 63, 64, 235, 261
Standpipes and piezometers 70-72
Static cones 63, 65, 66
Stepped face working 206
Stone columns 151
Stresses beneath foundations 245-258
Stress-strain behaviour 4
Structural defects 207, 208
Structure of soil 111, 113
Sulphate content test 103
Sulphates 128, 160
Surcharging 150, 151
Surface slides 309-312
Swelling/shrinkage problems 3, 5
Subgrade protection 3
Subsidence due to mining 165-169, 198-203, 205-206, 208
Swelling potential 144

Telephone services 41
Terrain mapping 42
Test results, use in design 36, 77-79
Tests
  see Laboratory, Field, In-situ, etc.
Tie rods 33, 34
Total and effective stresses 223-226
Trial pits
  see Boreholes and trial pits
Triaxial tests 86, 126, 220, 235,
  consolidated 79, 89
  drained 79, 88
  quick undrained 79, 87, 139
Tropical residual soils 108, 160

U100/U4 sampler 56, 59-61
Ultimate bearing capacity
  foundations 4, 126, 127, 137, 138, 227, 233-241, 244, 245
  rigid retaining walls 296
Unified classification system 118
Uplift 24, 25

Vane shear test 79
Variability of ground conditions 6, 12
Vertical drains 151
Vibro-compaction 151

Void migration 166, 179, 183
Voids ratio 217
Volcanic soils 108, 109, 160

Water content
  see Moisture content
Water pressure behind retaining walls
    139
Water replacement density test 97, 98
Water supply 41
Weathering grades 116, 117
Wedge analysis
  retaining walls 286
  slopes 312-315

Young's modulus 252

Zero air voids line 95